Springer Theses

Recognizing Outstanding Ph.D. Research



For further volumes:
http://www.springer.com/series/8790

## Aims and Scope

The series "Springer Theses" brings together a selection of the very best Ph.D. theses from around the world and across the physical sciences. Nominated and endorsed by two recognized specialists, each published volume has been selected for its scientific excellence and the high impact of its contents for the pertinent field of research. For greater accessibility to non-specialists, the published versions include an extended introduction, as well as a foreword by the student's supervisor explaining the special relevance of the work for the field. As a whole, the series will provide a valuable resource both for newcomers to the research fields described, and for other scientists seeking detailed background information on special questions. Finally, it provides an accredited documentation of the valuable contributions made by today's younger generation of scientists.

## Theses are accepted into the series by invited nomination only and must fulfill all of the following criteria

- They must be written in good English.
- The topic should fall within the confines of Chemistry, Physics, Earth Sciences, Engineering and related interdisciplinary fields such as Materials, Nanoscience, Chemical Engineering, Complex Systems and Biophysics.
- The work reported in the thesis must represent a significant scientific advance.
- If the thesis includes previously published material, permission to reproduce this must be gained from the respective copyright holder.
- They must have been examined and passed during the 12 months prior to nomination.
- Each thesis should include a foreword by the supervisor outlining the significance of its content.
- The theses should have a clearly defined structure including an introduction accessible to scientists not expert in that particular field.

Yue Yanan

# How Free Cationic Polymer Chains Promote Gene Transfection

Doctoral Thesis accepted by
The Chinese University of Hong Kong

*Author*
Dr. Yue Yanan
University of Chicago
Chicago
USA

*Supervisor*
Prof. Dr. Chi Wu
The Chinese University of Hong Kong
Hong Kong
People's Republic of China

ISSN 2190-5053 ISSN 2190-5061 (electronic)
ISBN 978-3-319-34378-5 ISBN 978-3-319-00336-8 (eBook)
DOI 10.1007/978-3-319-00336-8
Springer Cham Heidelberg New York Dordrecht London


Softcover reprint of the hardcover 1st edition 2013


Printed on acid-free paper

Springer is part of Springer Science+Business Media (www.springer.com)

**Parts of this thesis have been published in the following journal articles: (*: corresponding author; †: co-first author):**

1. **Y. Yue***, C. Wu*, Development of non-viral polymeric vectors for gene transfection, ***Biomaterials Sci***. Invited review, 1 (2) (2013), 152–170.
2. **Y. Yue***, F. Jin, R. Deng, J. Cai, Z. Dai, M. Lin, H. Kung, C. Wu*, Revisit complexation between DNA and polyethylenimine – Effect of length of free polycationic chains on gene transfection, ***J. Control. Release*** 152 (1) (2011), 143–151.
3. **Y. Yue***, F. Jin, R. Deng, J. Cai, Y. Chen, M. Lin, H. Kung, C. Wu*, Revisit complexation between DNA and polyethylenimine – Effect of uncomplexed chains free in the solution mixture on gene transfection, ***J. Control. Release*** 155 (1) (2011), 67–76.
4. R. Deng*, **Y. Yue**, F. Jin, Y. Chen, H. Kung, M. Lin, C. Wu, Revisit the complexation of PEI and DNA – How to make low cytotoxic and high efficient PEI gene transfection non-viral vectors with a controllable chain length and structure? ***J. Control. Release*** 140 (1) (2009), 40–46. (**Cover Story**)
5. J. Cai, **Y. Yue**, R. Deng, Y. Zhang, S. Liu, C. Wu*, Effect of chain length on cytotoxicity and endocytosis of cationic polymers, ***Macromolecules*** 44 (7) (2011) 2050–2057.
6. J. Wang, **Y. Yue**, G. Chen, J. Xia*, Protease-promoted drug delivery using peptide-functionalized gold nanoparticles, ***Soft Matter*** 7 (16) (2011) 7217–7222.
7. R. Deng, S. Diao, **Y. Yue**, T. Ngai, C. Wu, F. Jin*, Dynamic and structural scalings of the complexation between *p*DNA and *b*PEI in semidilute and low-salt solutions, ***biopolymers*** 93 (6) (2010), 571–577.
8. **Y. Yue**†, J. Cai†, C. Wu*, Quantitative three-dimensional analysis of endocytosis and intracellular trafficking of DNA/polymer complexes without/with free polycationic chains using confocal laser scanning microscopy, To be submitted.

# Supervisor's Foreword

This Ph.D. thesis has unearthed one critical and important fact which has been overlooked for a few decades in the development of non-viral vectors for gene delivery; namely, it is those uncomplexed cationic polymer chains free in the solution mixture of polymers and DNA that facilitate and promote gene transfection. By using a combination of synthetic chemistry, polymer physics, and molecular biology, Dr. Yue has confirmed that it is those cationic polymer chains free in the solution mixture, rather than those bound to DNA chains, that play a decisive role in the intracellular space.

Instead of the previously proposed and widely accepted "proton sponge" model, we proposed a new hypothesis based on the results of several well-designed and decisive experiments, i.e., free polycationic chains with a length longer than $\sim$10 nm are able to partially block the fusion between different endocytic vesicles, including the endocytic-vesicle-to-endolysosome pathway, so that the internalized polyplexes have a higher chance of being transported into the nucleus.

This thesis study is highly original and its results greatly deepen our understanding in polymer-mediated gene transfection. It is not an exaggeration to state that this thesis is paradigm shifting. I am delighted and feel lucky to have had Dr. Yue as one of my Ph.D. students because she was a highly self-motivated student. She majored in polymer science in her undergraduate studies and independently learned cell- and molecular biology as a postgraduate student. It was she who taught me some methods in molecular biology.

This thesis opens several doors for further studies. For example, how interfering snare proteins affect the intervesicular fusion, a frontier in molecular biology by itself; the detailed path of how the polyplexes are translocated into nucleus; and where the gene is released, inside or outside of nucleus. In summary, this thesis provides a new insight for how to rationally design a next generation of superior polymeric non-viral gene delivery vectors.

Hong Kong, April 2013 Prof. Dr. Chi Wu

# Acknowledgments

First of all, I would like to express my sincere thanks to my supervisor, Prof. Chi Wu, who gave me the precious opportunity to initiate an interdisciplinary project based on polymer physics, chemistry, and cell biology. I have not only benefited much from his rigorous trainings but also deeply impressed by his unrestrained motivation and diligence for scientific research.

My sincere thanks are also given to Prof. Kung Hsiang-fu, who led me into the mysterious biology world. His insightful guidance and encouragement, together with the sound trainings in basic theories and experimental skills would be of great help in my future.

I am also grateful to my dear group members for their stimulating discussions and generous help in daily life. They are Dr. Jin Fan, Dr. Deng Rui, Dr. Hong Liangzhi, Dr. Diao Shu, Ms. Cai Jinge, Mr. Ma Yongzheng, Ms. Dai Zhuojun, Mr. Zhao Hong, Mr. Chen Qianjin, Dr. Ge Hui, Dr. Gong xiangjun, Mr. Wang Jianqi, and Mr. Li Yuan. I really appreciate their efforts in my project.

My sincere gratitude is also extended to all the staff members in the Department of Chemistry, The Chinese University of Hong Kong for the active academic atmosphere and the service they offered. This research project would not be possible without the financial support from the RGC Earmarked Grant (CUHK4037/06P, 2160298). The generous supports are gratefully acknowledged.

Last but not least, I would like to give my sincere thanks to my family and Liu Jianzhao for their emotional supports during my pursuit of a higher degree. Thank you so much!

Hong Kong, March 2012 Dr. Yue Yanan

# Contents

# Abstract

The development of safe, efficient, and controllable gene-delivery vectors has become a bottleneck to human gene therapy. Synthetic polymeric vectors, although safer than their viral counterparts, generally do not possess the required efficacy. Currently, the exact mechanisms of how these polymeric vectors navigate to pass each intracellular gene delivery obstacle ("slit") and their particular physical/chemical properties that contribute to the efficient intracellular trafficking remain largely unknown, making it rather difficult to further improve the efficacy of non-viral polymeric vectors in vitro and in vivo.

The complexation and condensation of long anionic DNA with cationic polymer chains into small aggregates ($\sim 10^2$ nm) is the first and necessary step in the development of polymeric vectors. Our revisit of the complexation between DNA and polyethylenimine (PEI), one of the most efficient cationic polymeric vectors, by using a combination of laser light scattering and gel electrophoresis demonstrates that nearly all the DNA chains are complexed with PEI to form polyplexes when the molar ratio of nitrogen from polymer to phosphate from DNA (N:P) reaches ~3, irrespective of the chain length of PEI and solvent used. However, a high in vitro gene transfection efficiency is only achieved when N:P $\geq$ 10. Putting these two facts together, it has been concluded that (1) each solution mixture with a higher N:P ratio actually contains two kinds of cationic chains: bound to DNA and free in the solution (~70 %); and (2) it is those free PEI chains that actually promote the gene transfection no matter whether they exist (are added) many hours before or after the administration of the polyplexes (N:P = 3).

Further, effects of the length and topology of both the bound and free polycationic chains on the gene transfection were, respectively, studied. Notably, both short (~2 K) and long (~25 K) PEI chains are capable of condensing DNA completely at N:P ~ 3, but long free chains are $\sim 10^2$-fold more effective in enhancing the gene transfection, indicating that the length of free chains plays a vital role in the gene transfection. It is also interesting to note that for long free PEIs, the chain topology has nearly no effect on the transfection efficiency; but for short PEIs, linear free chains are ~10-fold more effective than branched ones. These results illustrate that bound PEI chains mainly play a role in DNA condensation and protection, while it is those polycationic chains free in the solution mixture that should get our attention.

To address how free polycationic chains with a proper length/topology promote the gene transfection, we quantitatively compared the cellular internalization and lysosomal distribution of polyplexes in the presence of different free PEIs by using a confocal image-assisted three-dimensionally integrated quantification (CIQ) method. Cellular uptake kinetics reveals that long free PEI chains boost the uptake rate, but their major contribution is in the intracellular space. In the clathrin-mediated (CME) pathway, the efficacy of free cationic chains is tightly correlated to their ability in preventing the entrapment of polyplexes into the late endosomes/ lysosomes. Namely, without free PEI chains or with short branched ones, the fraction of polyplexes inside lysosomes keeps escalating and reaches ~40 % in the first 6 h; whereas with short linear or long free chains, the percentage of polyplexes trapped into lysosomes is reduced to less than ~15 %. Notably, the well-accepted "proton sponge" effect plays a certain, but not dominant and decisive, role here because the shut down of proton pump only partially attenuates the gene transfection efficiency.

In addition to the CME pathway, the efficacy of long cationic free PEI chains in promoting the intracellular trafficking is also revealed in the caveolae-mediated pathway. At N:P = 4, with limited amount of free PEIs, the gene transfection was almost completely abolished when the caveolae-mediated pathway was blocked, whereas the gene transfection efficiency was enhanced by ~3 times when the CME pathway was inhibited. This indicates that at lower N:P ratios (≤4), most of the polyplexes are destined to the acidic lysosomes in the CME pathway, and only the caveolae-dependent route leads to an effective gene transfection. On the other hand, at N:P = 7, with sufficient amount of long free PEIs, the inhibition of either the CME or caveolae route decreased the transfection efficiency only by ~2 times. Collectively, our study found that free cationic PEI chain with a proper length (15–20 nm) are able to facilitate the intracellular trafficking, presumably via (1) blocking the signal proteins on the inner cell membrane so that the endolysosomes development is prevented or slowed down; and (2) compromising/disrupting the membrane of the virginal endocytic vesicles, early endosomes or caveosomes so that polyplexes entrapped inside can escape into the cytosol.

# Symbols and Abbreviations

|  |  |
|---|---|
| CD | Cyclodextrin |
| CDP | Cyclodextrin-based polycation |
| CIQ | Confocal image-assisted three-dimensionally integrated quantification |
| CLSM | Confocal laser scanning microscopy |
| CME | Clathrin-mediated endocytosis |
| CONTIN | A regularization program from Provencher |
| $D$ | Translational diffusion coefficient |
| $dn/dC$ | Differential refractive index increment |
| EGFP | Enhanced green fluorescence protein |
| FBS | Fetal bovine serum |
| FITC | Fluorescein-5-isothiocyanate |
| $f(R_h)$ | Hydrodynamic radius distribution |
| $G^{(2)}(\tau)$ | Intensity–intensity time correlation function |
| $<I(q)>$ | Time-averaged scattering intensity at a finite scattering angle |
| $<I(0)>$ | Time-averaged scattering intensity when the scattering angle → 0 |
| LDH assay | Lactate dehydrogenase assay |
| LLS | Laser light scattering |
| $M_n$ | Apparent number-averaged molar mass |
| $M_w$ | Apparent weight-averaged molar mass |
| MTT | 3-(4,5-dimethylthiazol-2-yl)-2,5-diphenyltetrazolium bromide |
| N:P | Molar ratio of nitrogen from polymer to phosphate from DNA |
| PBS | Phosphate buffered saline |
| *p*DNA | Plasmid DNA |
| PEI | Polyethylenimine |
| *b*PEI-0.8K | Branched PEI with a weight-averaged molar mass of 800 g/mol |
| *b*PEI-2K | Branched PEI with a weight-averaged molar mass of 2,000 g/mol |
| *b*PEI-25K | Branched PEI with a weight-averaged molar mass of 25,000 g/mol |
| *lin*PEI-2.5K | Linear PEI with a weight-averaged molar mass of 2,500 g/mol |

| | |
|---|---|
| *lin*PEI-25K | Linear PEI with a weight-averaged molar mass of 25,000 g/mol |
| PDMAEMA | Poly[2-(dimethylamino) ethyl methacrylate] |
| PEG | Polyethylene glycol |
| PLL | poly(L-lysine) |
| $T_f$ | Transferrin |
| $q$ | Scattering vector |
| $<R_g>$ | Average radius of gyration |
| $<R_h>$ | Average hydrodynamic radius |
| RLU/mg protein | Relative luminescence unit per cellular protein |
| $\zeta_{potential}$ | Zeta-potential |
| $\theta$ | Scattering angle |
| $<\rho>$ | Average chain density |

# Chapter 1
# Introduction and Background

Gene therapy, considered as treating genetically-caused diseases by transferring exogenous nucleic acids into specific cells of patients, has attracted great interests over the past few decades [1]. Advances in molecular biology and biotechnology as well as the completion of the Human Genome Project have led to the recognition of numerous diseases-relating genes [2]. It has been gradually and generally realized that development of safe, efficient and controllable gene-delivery vectors is now a bottleneck in clinical applications [2, 3].

The gene delivery vectors can be generally divided as viral and non-viral ones. Virus, such as adenovirus, adeno-associated virus, lentivirus and retrovirus, has evolved as a sophisticated gene-delivery vehicle and can be readily transformed into a viral vector by replacing part of its genome with a therapeutic gene [2]. Such recombinant viral vectors are efficient but at the same time potentially dangerous, previously leading to severe immune/inflammatory reactions or even cancer in patients [4–6]. Non-viral vectors, including synthetic polymers and lipids, offer some advantages over their viral counterparts, such as low immune toxicity, construction flexibility and facile fabrication [2, 7]. In particular, cationic polymers have attracted much interest because it is relatively easy to tune their chemical and physical properties through polymer chemistry so that they can acquire multiple functions for gene delivery. However, at this moment, their low gene transfection efficiency has greatly limits their clinical applications. For instance, polyethylenimine (PEI), one of the few most effective and versatile polymeric vectors, still remains $\sim 10^5$ times less efficient than its viral counterpart [8]. Since the first demonstration of polycation-mediated gene transfection in 1987 [9], hundreds, if not thousands, of cationic polymers with different chain lengths and topologies have been synthesized and explored as non-viral vectors for gene delivery.

Among them, PEI is the most intensively studied example [10, 11] and hitherto exhibits nearly the highest in vitro transfection efficiency in the absence of any exogenous endosomolytic agent. The optimal efficacy of PEI has been attributed to its unique ability to navigate through a series of intracellular "slits", including the escape from lysosomes (acidic vesicles filled with degradative enzymes), nuclear

Y. Yanan, *How Free Cationic Polymer Chains Promote Gene Transfection*,
Springer Theses, DOI: 10.1007/978-3-319-00336-8_1,

localization and DNA unloading [12–16]. However, the exact mechanisms of how these cationic chains overcome each intracellular gap and their particular physical/chemical properties that contribute to the efficient intracellular trafficking, remain largely unexplained, making it rather difficult to further improve the in vitro and in vivo efficiency of synthetic polymer-based non-viral vectors.

In this context, deciphering intracellular trafficking mechanisms and establishing the structure–function relationship of polymeric non-viral vectors are of great importance because they will rationally guide our design and create multipotent gene-delivery non-viral vectors for biomedical applications. There have been a number of review articles and chapters that summarize the recent developments in preparation and applications of non-viral polymeric vectors for gene transfection. In this introduction, we will not cover all the aspects in the field of polymer-mediated nucleic acid delivery. Instead, we will first give a brief overview of the synthetic polymeric vectors, and in particular, highlight some promising candidates for clinical applications. Our main focus is on summarizing and discussing some new insights on the intracellular trafficking mechanisms of DNA/polymer complexes ("polyplexes"). Particularly, we will identify and discuss four critical, but often overlooked, issues for successful DNA/polymer intracellular trafficking; namely, (1) effect of the uncomplexed polycationic chains free in the DNA and polymer solution mixture on the gene transfection; (2) effect of the endocytosis pathway on the intracellular fate of the polyplexes; (3) effect of the so-called and well-accepted "proton sponge" concept; and (4) effect of the nuclear localization and unloading of DNA inside the nucleus.

## 1.1 Non-Viral Vectors Made of Commercial and Specially Designed Polymers

For understandable reasons, many earlier gene-delivery studies used commercially available polymers, such as poly(L-lysine) (PLL) [9, 17], polyethylenimine (PEI) [10] and polyamidoamine (PAMAM) dendrimers [18]. These off-the-shelf polymers have been intensively studied and formed a literature basis for the development of non-viral gene delivery. In recent years, many types of novel polymers have been specifically designed and synthesized for gene delivery. In most of the cases, they were designed to pass one or more particular extra-/intracellular obstacles ("slits"), such as stability for systematic administration, release from endolysosomes and localization inside nucleus. Some of these polymers perform better than the best off-the-shelf polymers. However, none of them are able to rival viral vectors in clinical applications because of their lower gene transfection efficiency and higher cytotoxicity, especially when long polycationic chains are used for in vivo studies.

In general, polymers designed and explored for gene delivery include: (1) polyethylenimine (PEI) and its derivatives; (2) polymethacrylate; (3) carbohydrate-based polymers, generally with β-cyclodextrin, chitosan, dextran, poly(glycoamidoamine) and schizophyllan as their carbohydrate functionalities; (4) poly(L-lysine) (PLL); (5) linear poly(amido-amine) (PAA); (6) dendrimer-based vectors, such as PAMAM and poly(propylenimine) dendrimers; (7) biodegradable polymers, including phosphorus-containing polymers, poly(amino-ester), poly(4-hydroxy-L-proline ester), poly[α-(4-aminobutyl)-L-glycolic acid] (PAGA); and (8) polypeptide vectors, such as Tat-based peptide, antennapedia homeodomain peptide and many other examples reviewed in Ref. [19]. To illustrate their specific advantages and disadvantages, we choose to review several important classes of cationic polymers and emphasize their promising clinical trials as follows.

### *1.1.1 Polyethylenimine (PEI)*

The introduction of PEI as a non-viral vector represented a big leap because of its much higher efficiency in promoting the gene transfection compared to other early polymeric vehicles (e.g., PLL) [10, 11]. PEI mainly has two different topologies: linear and branched structures. Branched PEI (*b*PEI) is synthesized via acid-catalyzed polymerization of aziridine [20], whereas linear PEI (*l*PEI) is normally made by ring opening polymerization of 2-ethyl-2-oxazoline followed by hydrolysis (Scheme 1.1) [21]. Several *l*PEIs have been made as commercial transfection agents, including ExGen500 and jetPEI. Both of them are derivatives of *l*PEI with a molar mass of 22,000 g/mol.

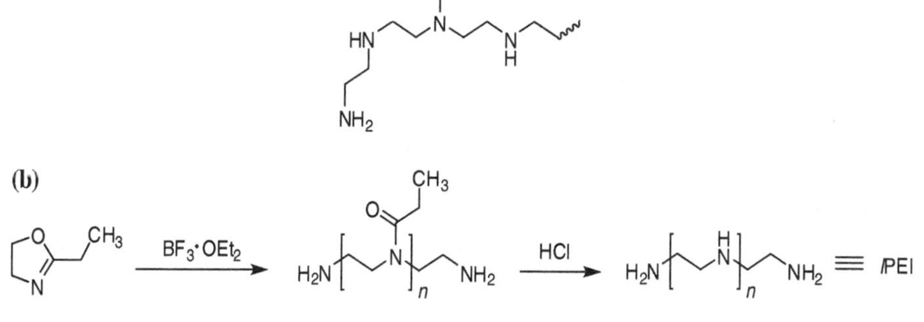

**Scheme 1.1** Schematic of synthesis of PEI by **a** branched: acid-polymerization of aziridine; and **b** linear: ring-opening polymerization of 2-ethyl-2-oxazoline followed by hydrolysis

Scheme 1.1 shows that PEI contains nitrogen at every third atom, leading to a high charge density on the chain, especially in acidic condition. Theoretical calculation shows that *b*PEI contains primary, secondary and tertiary amino groups with a 1:2:1 ratio [22]. These amines have $pK_a$ values spanning the physiological pH range, acted as a buffer. The degree of protonation of these amines increases from $\sim$20 to $\sim$45 % as pH decreases from $\sim$7.4 to $\sim$5.0 [23]. Previous studies have attributed its high transfection efficiency to the so-called "proton sponge" effect. Namely, further protonation of PEI chains inside the endolysosomes would lead to an influx of counter (chloride) ions and increase the osmotic pressure inside, which could burst the endocytic vesicle and release the polyplexes [10, 24]. Many people, especially those who jumped into this research field later, have taken such an explanation as granted. However, a number of researchers in the field have always questioned whether such a "proton sponge" effect plays a dominant role in promoting the gene transfection because of some realistic estimation of the additional osmotic pressure and some contradictive results [8], which will be discussed later.

It is well-known in the field that both the gene transfection efficiency and toxicity of PEI are strongly related to its chain length and topology (branched or linear) [25–28]. Long PEI chains are highly effective but more cytotoxic. It has been shown that free cationic PEI chains, in particular long ones (*b*PEI with a molar mass of 25,000 g/mol, denoted as *b*PEI-25K), can induce the membrane damage (necrotic-like changes) in the early stage (0.5 h), assessed by a considerable release of lactate dehydrogenase (LDH) from the intracellular to the extracellular space; while in the later stage (24 h), they initiate mitochondrial-mediated apoptosis, reflected by the release of cytochrome c, a subsequent activation of executioner caspase-3, and an alteration in mitochondrial membrane potential (MMP) [29, 30]. In addition to necrosis and apoptosis, it was recently shown that autophagy is also associated with the PEI-mediated cytotoxicity and gives rise to aggravated cell damage. Particularly, in the early stage ($\sim$3 h), autophagy is mainly correlated to the lysosomal damage, while in the later stage (after a 24-h recovery), autophagy is primarily associated with the mitochondrial injury [31]. Note that PEI is just one example. Many other polycationic chains also suffer from this "malignant" correlation of efficacy and toxicity. How to solve such a catch-22 problem has puzzled researchers for years. The search for a high efficient and low cytotoxic polymeric vector is an endless endeavor.

One of the approaches to circumvent such a catch-22 problem is to link short PEI chains into a long one by some degradable coupling agents, i.e. ester, $\beta$-aminoester and disulfide, for the development of high efficient and low cytotoxic non-viral vectors in recent years [32–36]. The former two linkers have a hydrolysis half-life time ranging from hours to days, which might not be sufficiently fast. In contrast, the reductive degradation of a disulfide is more quick in the presence of glutathione (GSH) in the cytosol [37, 38]. Goepferich et al. [34] clearly confirmed the intracellular degradation of such disulfide cross-linked PEI chains and the release of nearly non-toxic short PEI fragments (cell survival 98.69 ± 4.79 %) (Fig. 1.1a). In comparison with seven commercial transfection agents in seven different cell lines, these disulfide cross-linked *l*PEI vectors exhibited a superior efficiency and a substantially lower toxicity, showing that the reductive degradation is promising in designing new non-viral vectors (Fig. 1.1b).

Recently, Deng et al. [39] developed a laser light scattering (LLS) method to in situ monitor the disulfide-coupling reaction under a programmable mixing of short *b*PEIs ($M_w = 2 \times 10^3$ g/mol) and dithiobis(succinimidyl propionate) (DSP). In this way, a series of linked PEI chains with desired different molar masses can be obtained from one reaction mixture. A comparative study of the transfection

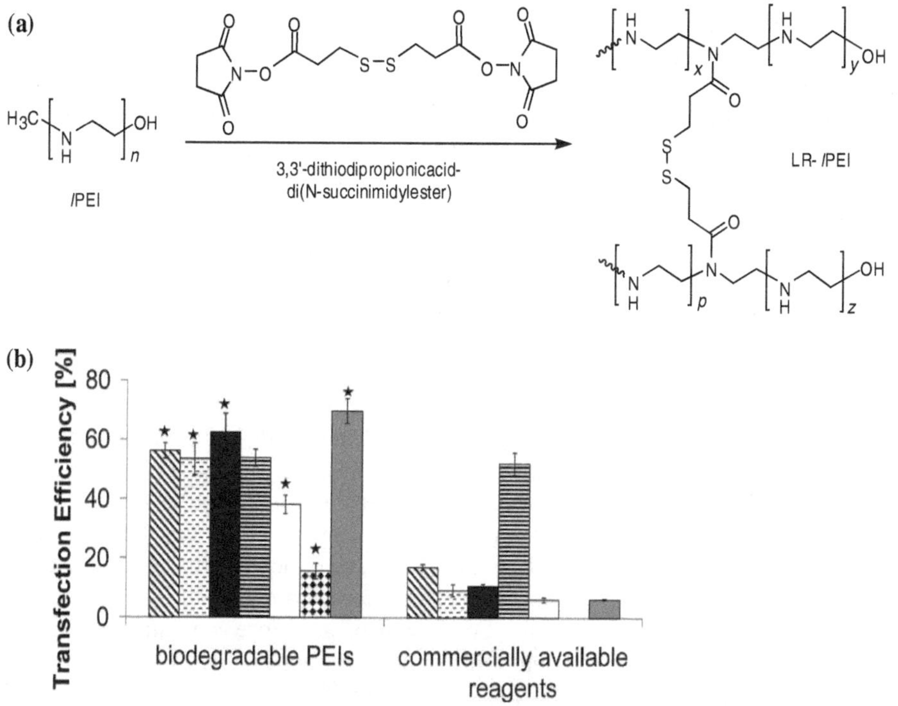

**Fig. 1.1** **a** Synthesis of disulfide cross-linked low-molecular-weight *l*PEI, where LR refers to Lomant's reagent, a cross-linking reagent. **b** Comparison of the maximum gene transfection efficiency of seven disulfide cross-linked *l*PEIs with seven commercially available transfection agents under conditions where the cell viability is >90 %, in (from *left* to *right*) CHO-K1, COS-7, NIH/3T3, HepG2, HCT116, HeLa and HEK-293 cells, respectively. Statistically significant differences of biodegradable PEIs compared with the commercial transfection agents are denoted by *filled star* ($p < 0.01$) (reprinted from Ref. [34] with permission of National Academy of Sciences, USA)

activity and cytotoxicity of two such linked PEI samples (PEI-7K-L and PEI-400K-L, respectively with $M_w = 6.5 \times 10^3$ and $3.8 \times 10^5$ g/mol) reveals that PEI-7K-L with an extended chain structure is less cytotoxic and 2–10 times more efficient than both the "golden standard" *b*PEI-25K and the widely-used commercial Lipofectamine 2000. On the other hand, PEI-400K-L with a microgel structure is ineffective in spite that it is much less cytotoxic. This study clearly demonstrates that a proper control of the chain structure is more important than that of the overall molar mass.

Previous studies have shown that the "naked" PEI-based polyplexes, although possess positive surface charges, tend to aggregate in a time-dependent manner in physiological buffers (ionic strength equals to that of 150 mM NaCl) [40]. When administered in vivo, they are likely to absorb to the serum albumin and other negatively charged proteins in bloodstream, giving rise to further aggregation and a rapid clearance of them by phagocytic cells and the reticuloendothelial system (RES) [41]. To unravel such problems, the surface of the polyplexes was usually modified with a layer of hydrophilic polymers. When polyethylene glycol (PEG) is used, it is often called PEGylation. The steric and hydrophilic shell stabilizes the resultant polyplexes in the physiological condition, reduces their undesirable interaction with proteins, and also increases their intravenous circulation time [11, 42]. It should be noted that increasing the length and grafting density of PEG chains impedes the DNA complexation. Short PEG chains ($M_w \leq 500$ g/mol) fail to provide the shielding effect, while a molar mass of at least 2,000–5,000 g/mol seems to be sufficient to achieve such an effect [11]. Unfortunately, there is a dilemma about PEGylation because it makes the polyplexes more "stealth" in body but reduces the cellular internalization, hinders the intracellular unpacking, and hampers the following release of DNA in the nucleus [43, 44].

The attachment of properly chosen cell-targeting ligands at the end of each PEG chain can enhance the cellular uptake [11, 45, 46]. In clinical applications, it is often beneficial and sometimes critical to target the polyplexes to a specific cell type or tissue [2]. Over the past few decades, much effort has been done to conjugate targeting moieties to PEI chains to enhance their cellular uptake and cell specificity. Many receptor proteins on cell membrane are chosen for targeting via the receptor-mediated endocytosis. For instance, galactose was attached to PEI chains to target the asialoglycoprotein receptors on hepatocytes [47], while iron-transport protein transferrin ($T_f$) [46], folic acid [48, 49] and epidermal growth factor (EGF) [45, 50] were respectively conjugated to PEI chains to target their corresponding receptors that are typically up-regulated on cancer cells. The efficient cell-specific targeting requires careful optimization of various parameters, including the length of a spacer between ligand and polyplex, the number of ligands per polyplexes, and the ligand-receptor binding strength [2]. Notably, the attachment of proper targeting ligands to the periphery of the polyplexes not only improves their cellular internalization, but also alters their subsequent intracellular trafficking pathways [8, 49], as will be discussed later.

## 1.1.2 Poly[2-(dimethylamino) ethyl methacrylate] (PDMAEMA)

PDMAEMA bearing tertiary amino group in the side chain was utilized as a gene transfer agent in the early studies [51–53]. Narrowly distributed linear PDMAEMA chains with different desired lengths can be synthesized via the atom transfer radical polymerization (ATRP) (Scheme 1.2) [54]. Note that PDMAEMA is less effective in the gene transfection than PEI. The choice of PDMAEMA in study is mainly due to its well-documented synthesis and characterization so that it becomes an excellent model for the evaluation of relationships between the chain structures and functions [51, 52, 55]. Long et al. [55] found that the gene transfection efficiency was dramatically enhanced as the PDMAEMA chain became longer in the molar mass range of $4.3–92 \times 10^4$ g/mol, highlighting how significant the polycationic chain length is in the gene transfection. On the other hand, polyplexes made by different PDMAEMA chains showed a comparable high level of cellular uptake, clearly indicating that the intracellular trafficking, rather than the cellular entry, is the rate-limiting step in the PDMAEMA-mediated gene transfection.

In terms of cytotoxicity, Cai et al. [54] found that in the concentration range normally used for the in vitro gene transfection (10–110 μg/mL), PDMAEMA chains with different lengths are cytotoxic to HepG2 cells by different mechanisms. Namely, (1) for short PDMAEMA chains [$M_w = (1.1–1.7) \times 10^4$ g/mol], their cytotoxicity, membrane disruption, and apoptosis are very low, independent of the chain length; (2) in the medium range ($1.7 \times 10^4 < M_w < 3.9 \times 10^4$ g/mol), their cytotoxicity increases with the chain length and polymer concentration, mainly due to the cooperative effect of membrane disruption and apoptosis; and (3) long chains [$M_w = (3.9–4.8) \times 10^4$ g/mol] become much disruptive to cellular membrane and pro-apoptotic so that they are able to pass through the cytoplasm and enter the nucleus much faster than short chains but their high cytotoxicity is less dependent on the chain length.

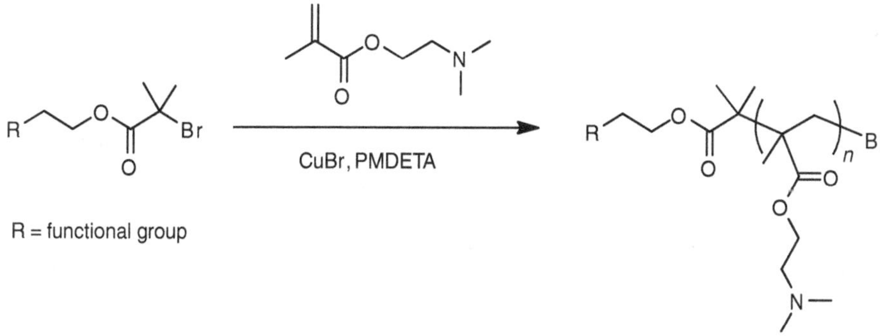

CuBr, PMDETA

R = functional group

**Scheme 1.2** Synthesis of linear PDMAEMA with different desired chain lengths via atom transfer radical polymerization (ATRP)

## 1.1.3 Cyclodextrin-Based Polymers

Cyclodextrins (CDs), cyclic oligosaccharides made of 6, 7 or 8 glucose units (called α-, β- and γ-CD, respectively) are pharmaceutically attractive; namely, (1) they can form water-soluble inclusion complexes with small, hydrophobic "guest" molecules, e.g., adamantine (AD); and (2) they elicit no immune responses and have very low cytotoxicity so that they are approved by FDA as solubilizing agents in pharmaceutical formulations [56]. In 1999 Davis et al. [57] first incorporated β-CD into the backbone of linear polycationic chains to introduce a new class of CD-based gene delivery vectors (Fig. 1.2a). The initial study showed that these CD-containing polymers are not only as effective as PEI and Lipofectamine, but also have minimal toxicity in both BHK-21 and CHO-K1 cell lines. The effect of CD size, charge centre and charge density on the gene-delivery efficacy and polymer toxicity were explored and summarized in later publications, as reviewed in Ref. [58].

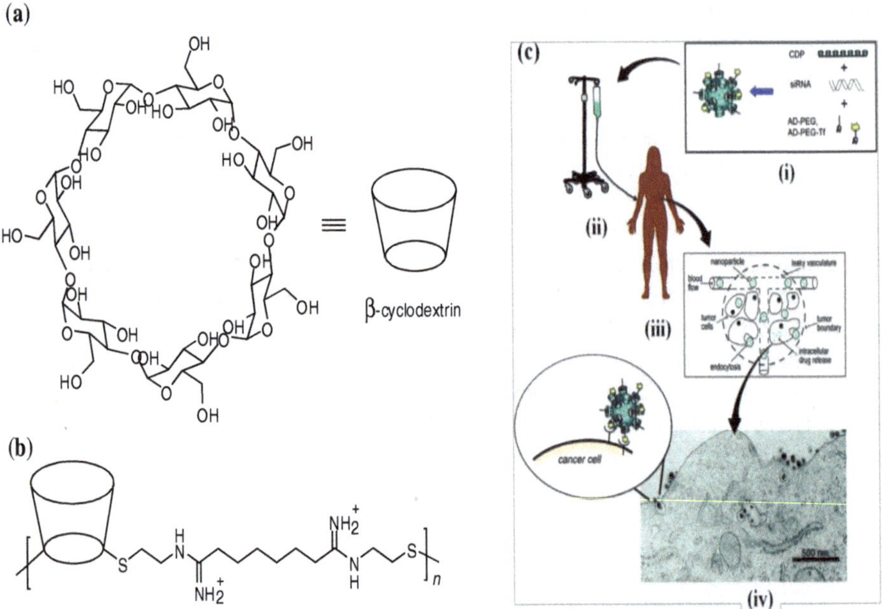


**Fig. 1.2** **a** Chemical structure of β-cyclodextrin (CD). β-CD has a hydrophobic interior and hydrophilic exterior surface. **b** Chemical structure of a β-CD-based polymer (CDP) designed for gene delivery, where n = 5. **c** Schematic of a targeted CDP-based nanoparticle delivery system made of a water-soluble CDP, an adamantine(AD)-PEG conjugate, a human transferrin conjugated at one end of PEG-AD for targeting, and siRNA, where an aqueous solution of nanoparticles is infused into patients, circulates in the blood, leaks via the effect of EPR into tumor tissues, penetrates though the tumor, and finally enters into the cancer cell via receptor-mediated endocytosis, as shown by a transmission electron micrograph (reprinted from Ref. [58] with ACS permission)

The CD-based polyplexes are readily to form inclusion complexes with some hydrophobic compounds. For example, they can be decorated with short AD-terminated PEG chains to improve their stability in biological fluids [59], or with AD-terminated galactose [59] and transferrin ($T_f$) [60] to target hepatocytes and cancer cells, respectively. This strategy has been used to a host of CD-based polymers [61], yielding a targeted delivery system made of a linear, CD-based polycation (CDP), a human $T_f$ protein ligand because $T_f$ receptors are typically up-regulated in cancer cells, a PEG steric stabilization agent, and plasmid DNA or small interfering RNA (siRNA) (Fig. 1.2b). Quickly, people realized that this kind of nanoparticles can be made by adding all the components together at one time. Namely, all the delivery components are placed in one vial; and DNA or siRNA, in another vial. Simply mixing them can lead to stable nanoparticles with a diameter of 60–80 nm, even at a very high nucleic acid concentration [62].

The CDP-based delivery system was successful in delivering plasmid DNA (*p*DNA) [63] and DNAzyme (short catalytic single-strand DNA) [64] to subcutaneous tumors via intravenous (i.v.) injection into mice. Later, it was shown that the $T_f$-targeted CDP-based nanoparticles with anti-tumor siRNA can significantly limit the proliferation in a disseminated murine model of Ewing's sarcoma [65]. A combination of bioluminescence imaging (BLI) and positron emission tomography (PET) further revealed the effect of tumor-specific targeting on their in vivo biodistribution and efficacy [66]. It should be noted that similar to the *p*DNA and DNAzyme deliveries [63, 64], those non-targeted siRNA nanoparticles are also able to accumulate in the tumor region through the effect of enhanced permeability and retention (EPR) but their internalization into tumor cells is much less efficient so that their carried genes could not be expressed [66]. Therefore, it should be reminded and also remembered that our primary objective of targeting is to enhance the uptake of the nanoparticles by tumor cells, rather than their accumulation around tumor cells, in spite that the accumulation is a necessary and important step for the cancer cell uptake, endocytosis.

The early in vivo success of this CDP-based targeted system motivated people to entail its translation from laboratory to clinic [58]. It has been shown that a targeted CDP-based delivery system, clinical version denoted as CALAA-01, is well-tolerated in multi-dosing experiments in a variety of non-human primates [67]. To our knowledge, Davis et al. [68] are now conducting the first in-human phase I clinical trial, which involves the systematic administration of siRNA therapeutics to patients with solid tumors by using this delivery system. Post-treatment tumor biopsies from melanoma patients showed that the amount of the nanoparticles localized in the intracellular space was correlated to the administrated dose level. Moreover, in comparison with the pre-dosing tissue, the reduction of both the messenger RNA (M2 subunit of ribonucleotide reductase (RRM2)) and the protein (RRM2) levels was detected [68].

### 1.1.4 pH-Sensitive, Membrane-Disruptive Polymers

Intracellular trafficking is critical to deliver a therapeutic gene because of its susceptible degradation in lysosomes by various enzymes. In nature, many viruses have evolved some specific acidic peptides in their protein coat that can be protonated at an acidic environment and thus become fusogenic with the endosomal membrane, allowing release of the therapeutic gene directly into the cytoplasm [69]. It motivated people to design and prepare a myriad of acid-stimulated membrane-disruptive polymers and hope that they can similarly facilitate the endosomal release as viral fusogenic peptides. Such endosomolytic polymers include both polyanions and polycations. Typically, a polyanionic endosomolytic system comprises (1) acid-responsive functionalities, especially –COOH and anhydride groups with a $pK_a$ in the endosomal range of 5.5–6.5; (2) hydrophobic groups to interact with and disrupt the endosomal membrane; (3) cationic pendant groups to complex/conjugate with a therapeutic gene; and (4) a tumor cell-targeting ligand [2]. One of the early examples is poly(2-ethylacrylic acid) (PEAA). It can undergo a hydrophilic-to-hydrophobic transition at $pH < 6$ so that it can partition into and disrupt the membrane of phospholipid vesicles [70, 71] and red blood cells [72]. Following this study, two related polymers, poly(2-propylacrylic acid) (PPAA) and poly(2-butylacrylic acid) (PBAA), were synthesized to examine whether making the pendant alkyl group more hydrophobic would increase the hemolytic activity, a reflection of the ability of agents to disrupt membranes [72, 73]. It was found that PPAA could disrupt the red blood cells 15-fold more efficiently than PEAA at pH $\sim$ 6, yet showed no hemolytic activity at pH $\sim$ 7.4. In contrast, PBAA led to a severe hemolysis even at physiological pH, making it undesirable for the development of non-viral vectors [73]. Inspired by its acid-responsive hemolytic activity, PPAA was incorporated into some cationic DNA/(1,2-dioleoyl-3-trimethylammonium-propane) (DOTAP) lipoplexes [74] and DNA/chitosan polyplexes [75], respectively; and remarkably improved their intracellular gene trafficking in vitro and in a murine excisional wound healing model [74–76].

The early success of PPAA and its derivatives motivated recent developments in making a family of modular diblock copolymers that are composed of a cationic block, PDMAEMA, to condense siRNA, and a second endosomolytic block comprising DMAEMA, 2-propylacrylic acid and butyl methacrylate (BMA), by using the controlled reversible addition fragmentation chain transfer (RAFT) polymerization [77]. Such diblock copolymers become sharply hemolytic at the endosomal pH regime and their hemolytic activity steadily increases with the BMA content in the second block. The siRNA-mediated knockdown of a model protein, glyceraldehyde 3-phosphate dehydrogenase (GAPDH), in HeLa cells was generally enhanced with the hemolytic activity of these pH-responsive copolymers [77]. Further, PPAA-based carriers was conjugated with a protein antigen, ovalbumin, via a disulfide bond to mediate its intracellular delivery in a mouse tumor model, potentially useful in vaccine applications [78].

In polycation category, an *N*-substituted poly(aspartamide) bearing 1,2-diaminoethane side chains [PAsp(DET)] exhibits minimal toxicity and great efficacy in mediating the release of polyplexes from the acidic endosomes due to its low-pH-stimulated membrane destabilization (Fig. 1.3a) [79, 80]. Similar to other endosomolytic agents, PAsp(DET) also manifests minimal membrane disruption at the physiological pH but becomes membrane-disruptive at the endosomal pH regime ($pK_a \sim 6.3$). Such a property is attributed to the protonation alteration in the flanking diamine unit, i.e., the monoprotonated gauche form at the neutral pH and the diprotonated anti form at the acidic pH, as shown in the inset of Fig. 1.3a. In other words, how 1,2-diaminoethane is protonated plays a pivotal role in the endosomal disruption [79]. It was also shown that the facile degradation of PAsp(DET), which is induced by a rapid self-catalytic reaction between the PAsp backbone and the side-chain amide nitrogen, minimizes the cumulative toxicity caused by polycationic chains [80]. Recently, Kataoka et al. [81] utilized a PEG-*b*-PAsp(DET) derivative together with intravital real-time confocal laser scanning microscopy (IVRTCLSM), for the first time, to in situ quantify the dynamic states of polyplexes in bloodstream and visually substantiate the efficacy of PEGylation in preventing the polyplexes agglomeration and polyplex-platelet interaction, as shown in Fig. 1.3b.

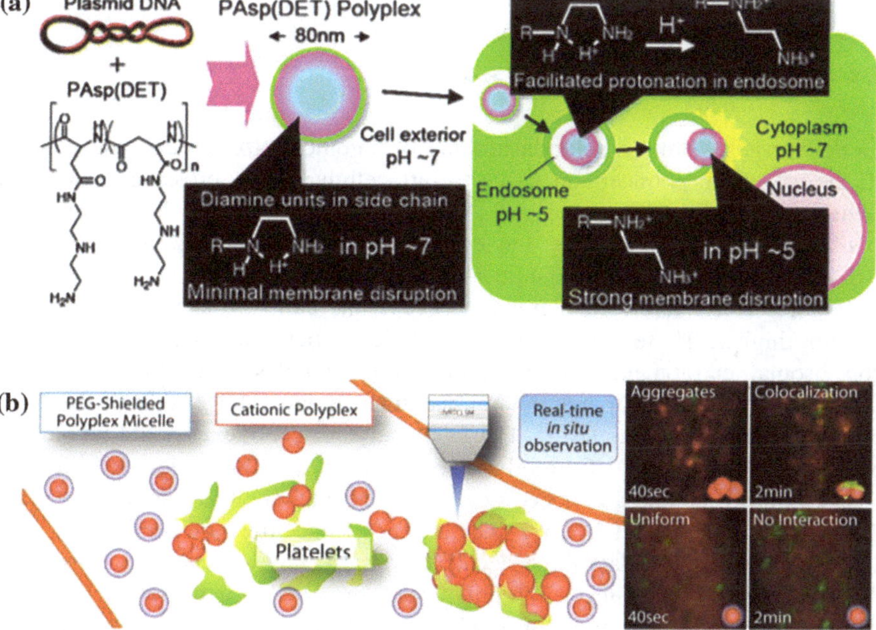


**Fig. 1.3** **a** Schematic of gene transfection mediated by PAsp(DET), where *inset* shows protonation of 1,2-diaminoethane moiety (reprinted from Ref. [79] with ACS permission); and **b** Schematic of interaction of polyplexes (*red*) with/without PEG coating with platelets (*green*) in bloodstream of mouse earlobe (reprinted from Ref. [81] with Elsevier permission)

Alternatively, Duncan et al. [82, 83] introduced a library of linear poly(aminoamine)s (PAA), which have a pH-dependent conformation and membrane perturbation ability, to deliver genes and protein drugs. The protonation of PAA reduces the freedom of chain conformation and leads to a more rigid chain structure. Such a conformational change in lower pH values enhances its hemolytic activity so that it can function as an endosomolytic agent [84]. Very recently, Richardson et al. [85] provided a direct evidence of how a PAA derivative (ISA1) in vivo permeabilizes the endocytic vesicular membranes, as shown in Fig. 1.4. In this study, radioactive-labeled ISA1 was combined with a liver sub-cellular fractionation to monitor the dose- and time-dependant passage of ISA1 along the endocytic pathway after an intravenous administration to rats, wherein the vesicular permeabilization (a reflection of the disruption of late endosomes/lysosomes) is quantified by the release of *N*-acetyl-D-glucosaminidase (NAG) from the vesicular fraction to the cytosolic fraction. The escalation of the ISA1 dose and incubation time enhances the release of both the radioactive polymer and NAG. Moreover, it was suggested that the endosomolytic activity of PAA might be due to the physical interaction between the PAA and membrane rather than the "proton sponge" effect. Notably, this study provides a general methodology to acquire "quantitative" information on the intracellular localization of polymeric vectors and their therapeutic cargos in vivo.

## 1.2 Important Remaining Issues

To deliver genes from a solution mixture of anionic DNA and cationic polymer all the way from extracellular space to intracellular space, crossing through the cellular membrane, the cytoplasm, and the nuclear membrane before releasing DNA inside the cell nucleus, the complexes made of polymeric vectors and DNA have to pass through a number of narrow gaps ("slits"), not "barriers" as widely described before in literature because one can pass a barrier as long as jumping higher. These "slits" mainly include endocytosis, escape from the endolysosomal entrapment, transportation inside cytoplasm, localization on and pass through the nuclear membrane, and final release of DNA from the polyplexes inside the nucleus [2, 8]. The DNA/polymer complexes could be blocked by any of these "slits". Currently, one of the most difficult issues is how to create a multi-functional delivery system so that the polyplexes are able to waltz through these "slits". It is astonished to know that after publishing more than 100,000 papers in the last three decades, we still have no clear pathway about the intracellular trafficking of the polyplexes, one of the first important steps in the development of effective non-viral vectors. In the following, we will discuss four critical and important, but often over-looked, issues on the intracellular trafficking of the DNA/polymer complexes.

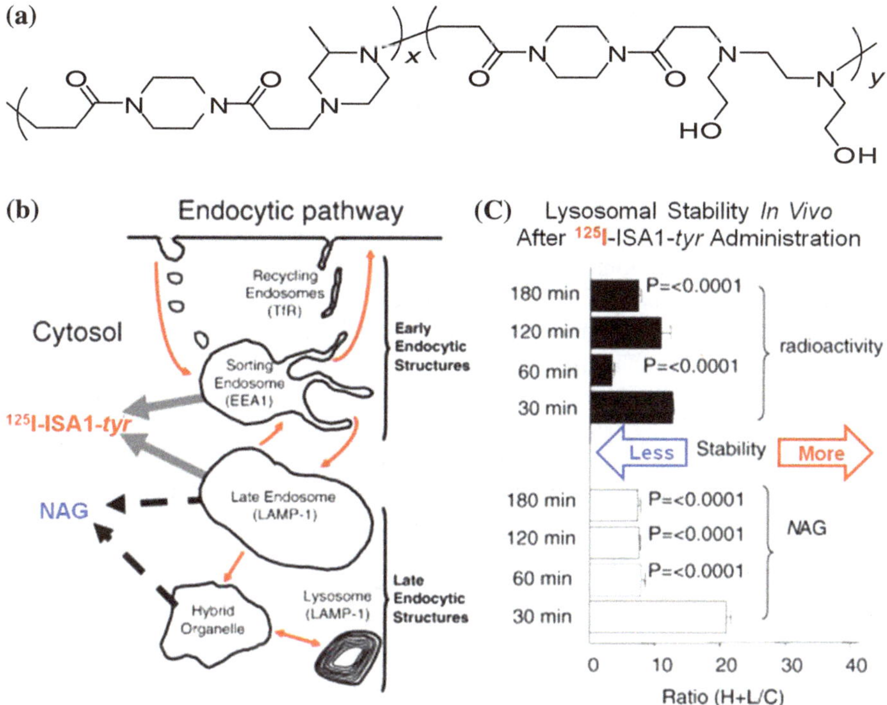


**Fig. 1.4** **a** Chemical structure of a poly(amidoamine) derivative, ISA1. **b** Schematic of endocytic system and markers used in sub-cellular fractionation studies; and **c** Time dependence of lysosomal stability index after the administration of [125] I-labelled ISA1-*tyr* at 10 mg/kg, where the lower the index, the greater the vesicle permeability (reprinted from Ref. [85] with Elsevier permission)

### *1.2.1 Role of Cationic Chains Free in Solution Mixtures of DNA and Polymer*

The complexation and condensation of long anionic DNA with cationic polymer chains into small aggregates ($\sim 10^2$ nm) is the first and necessary step in the development of non-viral polymeric vectors. It is worth noting that in literature the driving force of such complexation is often mistaken as electrostatic attraction; namely, an enthalpy driven process. Actually, it is driven by the gain of entropy, i.e., the release of small counter ions from both anionic DNA and polycationic chains during the complexation [86]. Due to the huge gain of translational entropy, the formation of DNA/polymer polyplexes is normally instantaneous and spontaneous upon the mixing of two aqueous solutions (DNA and polymer). Many efforts have been spent to correlate the size, density and surface charge of the polyplexes to their final gene transfection efficiency [26–28, 87], but there is no coherent picture. Previous studies revealed and confirmed two facts; namely, in order to achieve a reasonable transfection efficiency, (1) the periphery of the polyplexes in the solution mixture should be slightly positively charged; and (2) the molar ratio of nitrogen from polymer to phosphate from DNA (N:P) should be around 10. It is easy to understand that a positively charged periphery facilitates the attachment of the polyplexes to the negatively charged cellular membrane and endocytosis. For a long time, few people have asked why the N:P ratio has to be much higher than that required for the charge neutrality (N:P $\sim$ 1) [88, 89]. Only in 2004, Wagner et al. [90] found that a large amount of PEI chains are unbound to DNA and exist as individual chains free in the solution mixture. It was also found that these free cationic PEI chains are more toxic than those bound to DNA inside the polyplexes [90, 91]. Moreover, the removal of those free chains by size exclusion chromatography significantly reduced the gene transfection efficiency [90]. However, they did not follow up such a finding; namely, why and how those free polycationic chains help the gene transfection?

Only very recently, our group revisited the complexation between DNA and PEI in both water and phosphate buffered saline (PBS) by using a combination of laser light scattering and gel electrophoresis [92, 93]. We concluded that (1) each solution mixture with a higher N:P actually contains two kinds of cationic chains: bound to DNA and free in the solution ($\sim$70 %); and (2) it is those free PEI chains that actually promote the gene transfection no matter whether they exist (are added) many hours before or after the administration of the polyplexes (N:P = 3). This is the main focus of my Ph.D. thesis and will be discussed in details in following chapters.

### *1.2.2 Endocytosis Pathway on Intracellular Fate of Polyplexes*

Recently, different possible modes of polyplexes internalization have been correlated to their subsequent intracellular trafficking routes as well as the gene transfection efficiency [8, 94]. It is generally known that small polyplexes can be internalized by cells via multiple mechanisms, including clathrin-mediated endocytosis (CME, for endocytic vesicles with a size of $\sim$100–150 nm), caveolae-mediated endocytosis ($\sim$50–80 nm), micropinocytosis ($\sim$90 nm) and macropinocytosis ($\sim$500–2000 nm) [95–98].

In the CME pathway, the polyplexes are taken up by clathrin-coated pits, transferred to early/late endosomes and ultimately destined to lysosomes (Fig. 1.5a) [99]. Alternatively, small polyplexes can be internalized by caveolae, flask-shaped invaginations on the cell surface that bud from microdomains rich in cholesterol and caveolin, and subsequently delivered to caveosomes, pre-exsiting organelles with a stable neutral pH (Fig. 1.5b) [97–100]. The caveolae-mediated pathway might be more favorable for the gene transfection because there is a relatively less chance for caveolar vesicles to fuse with late endosomes or lysosomes [8, 101], presumably due to the lack of proper signal molecules required for inter-vesicular fusion [102]. Micropinocytosis initiates at the non-coated vesicles on the plasma membrane, which bud into the cytosol to form micropinosomes. Such non-coated vesicles become acidified and merge with early endosomes in common with the CME pathway [96, 98]. Macropinocytosis accompanies the actin-driven membrane ruffling which is regulated by growth factors or other signals. Such membrane protrusions collapse onto and fuse with the plasma membrane to generate large endocytic vesicles, called macropinosomes, that could engulf large polyplexes aggregates (> ~1 μm), as schematically shown in Fig. 1.5b [96, 98, 99]. Currently, the last three pinocytosis pathways remain poorly understood in comparison with the well-studied and documented CME pathway.

Previous studies showed that the internalization of the polyplexes made of "off-the-shelf" polymer, such as PEI and PDMAEMA, mainly follows the clathrin- and caveolae-mediated pathways [49, 103, 104]. Blocking of either one of them with a specific inhibitor only led to a partial, and sometimes, marginal decrease (<10 %) of the cell uptake of the polyplexes, indicating that these two uptake routes might be interchangeable [104]. On the other hand, the transfection efficiency was nearly

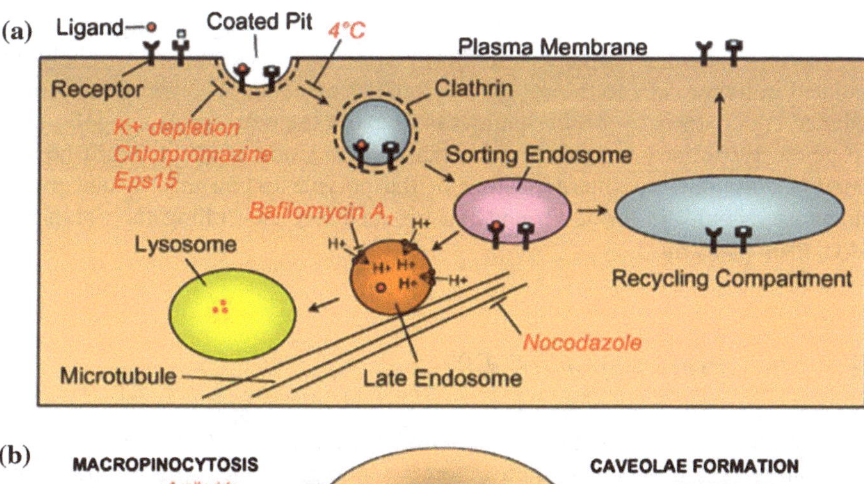


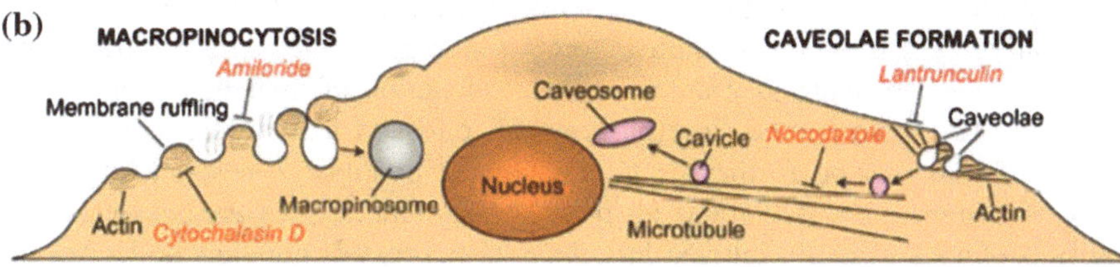


**Fig. 1.5** **a** Schematic of clathrin-mediated endocytosis, where internalized ligands are delivered either through a degradative pathway (leading to lysosomes) or a recycling pathway (leading to a recycle, back to the cell surface). **b** Schematic of macropinocytosis and caveolae- mediated endocytosis, where *red italics* delineate the inhibitors for indicated functions (reprinted from Ref. [99] with Nature permission)

completely lost when the caveolae-mediated pathway was blocked; but the transfection efficiency remained unchanged or even increased up to 2-fold after the CME pathway was inhibited in A549, HeLa [49, 103] and COS-7 cell lines [104]. Note that in these experiments the N:P ratio used was in the range 4–6; namely, free polycationic chains are limited in the solution mixture used. If the polyplexes were internalized merely via the CME pathway, such a small amount of free PEI chains might not be sufficient to prevent the entrapment of the polyplexes into acidic lysosomes [92, 93]. Therefore, at lower N:P ratios, the caveolae-dependent route is more likely to lead to an effective gene transfection.

Further, Pack et al. [49] investigated effects of two cell-targeting ligands, $T_f$ and folic acid, on the intracellular trafficking of the DNA/PEI polyplexes (N:P $\sim$ 4). It is known that $T_f$ and folic acid are typically internalized, respectively via the clathrin- and caveolae-mediated pathways. The attachments of these two ligands respectively to PEI chains via a covalent bond enable the successful delivery of their cargos through their respective pathways. Similar to previous results, the gene transfection efficiency was not adversely affected after the CME pathway was inhibited, but was entirely abolished after the caveolae-mediated pathway was blocked by small molecular drugs or RNA interference. It is further shown that targeting the polyplexes through the caveolae-mediated pathway prevents the rapid and direct fusion of small endocytic vesicles with more acidic late endosomes or lysosomes. These recent results reveal that an optimized targeting ligand for the gene therapy should 1) be able to associate with receptors that are typically up-regulated in tumor cells to improve the cellular uptake; and 2) favor the caveolae-mediated endocytosis over other pathways to avoid the delivery of the polyplexes into acidic lysosomes so that the enzymatic degradation of DNA could be prevented [8]. Meanwhile, internalization of the polyplexes via macro- and micro-pinocytosis should be further elucidated for precisely controlling the polyplexes endocytosis pathways.

### *1.2.3 The "Proton Sponge" Effect*

For polycationic chains with a proton buffer capacity, such as PEI and other pH-sensitive polymers, the so-called "proton sponge" effect on the intracellular trafficking is often taken as granted in the explanation of their high transfection efficiency. The heart of the "proton sponge" effect is that different amines on polymer chains can be further protonated inside endolysosomes, leading to an influx of counter ions ($Cl^-$) and an increase of osmotic pressure inside so that endolysosomes are finally bursted [10, 24]. However, this popular model has not been rigorously validated. Two fundamental issues related to this well-accepted model have to be considered. Firstly, the buffer capacity of a polymeric vector sometimes does not or even reversely correlate to the gene transfection efficiency. For instance, for a given topology, long PEI chains are more effective than short ones in the gene transfection [93]. If only considering their colligative properties, we know from thermodynamics that for a given weight concentration (g/mL), short chains should generate a higher osmotic pressure inside the endocytic vesicles. The previous results also showed that a simple coupling of 3–4 short *b*PEI-2K chains via a disulfide linker slightly decreases their buffer capacity but greatly enhances the gene transfection by a factor of $\sim 10^4$–$10^5$ times [39], depending on the N:P ratio, which could not be simply explained by the "proton-sponge" effect.

To investigate the structure-efficacy relationship, Thomas and Klibanov [105] performed a set of modifications of primary, secondary and tertiary amines on *b*PEI-25K and *b*PEI-2K, which decrease the number of protonable amines and lower their buffer capacities. Surprisingly, N-acylation of *b*PEI-25K with alanine nearly doubled its gene transfection efficacy in the presence of 10 % serum. Moreover, dodecylation and hexadecylation of primary amines on short *b*PEI-2K enhanced its gene transfection efficiency by $\sim$400 times in the presence of serum, even $\sim$5-fold higher than that of *b*PEI-25K. Further, Pack et al. [106] synthesized a series of modified PEI chains by acetylating different amounts of primary and secondary amines. Their results showed that partial acetylation reduces the buffer capacity, but increases the in vitro gene transfection efficiency of PEIs. Specifically, the acetylation of $\sim$43 % primary amines made PEI $\sim$ 20-fold more efficient than its parent (N:P $\sim$ 15–20) no matter whether serum was added. In opposite, Hennink et al. [107] attempted to improve the endosomolytic ability of PDMAEMA by introducing an additional tertiary amino group to each monomeric unit and hope to boost the "proton sponge" effect. Unexpectedly, such a modified PDMAEMA analogue exhibited much lower transfection efficiency even though it was less cytotoxic. However, adding an endosomolytic peptide, INF-7, restored the gene transfection efficiency, clearly indicating that the higher buffer capacity of modified PDMAEMAs are not able to mediate the polyplex release from endo-lysosomes via the "proton sponge" effect. In a similar way, Schacht et al. [108] showed that the imidazole-modified PDMAEMA derivatives had a similar buffer capacity as PEI but were still not able to transfect COS-1 cells, much worse than PDMAEMA with only tertiary amines. Further study revealed that these modified PDMAEMA chains were actually less effective in preventing the entrapment of the polyplexes into acidic late endosomes or lysosomes [109]. These studies do not completely deny a possible role of the buffer capacity played in the gene transfection. Instead, they emphasize that the buffer capacity should not play a dominating and decisive role in promoting the intracellular trafficking of polyplexes.

The second issue is whether the osmotic pressure generated by the "proton sponge" effect is sufficiently high to rupture the endocytic vesicle by itself or other mechanisms, such as polycation-membrane physical interaction at a lower pH, are also simultaneously involved. Assuming that one clathrin-coated vesicle contains one DNA/PEI complexes of N:P = 7, previous estimation showed that the maximum osmotic pressure produced inside this vesicle was $\sim 8.3 \times 10^4$ Pa even when pH was decreased from 7.4 to 5.0 [8]. Such a change in osmotic pressure would expand the membrane area only by 2.3 %, whereas lipid vesicles can generally withstand an area expansion up to ~5 % before they start to lose their integrity [110]. Note that in reality the proton-absorbing capacity of polycationic chains must be greatly attenuated because of their complexation with anionic DNA chains as well as their absorption to other anionic membranes and proteins. Therefore, the increase of osmotic pressure inside endocytic vesicles during the acidification alone is theoretically insufficient to rupture them, although it might be a cooperative factor to the eventual release of polyplexes from various intracellular vesicles [8, 15].

Besides a possible increase of osmotic pressure via the "proton sponge" effect, PEI can also destabilize the anionic membrane via the charge neutralization [13, 93], facilitating the endosomal release. Note that long *b*PEI-25K chains are much more disruptive to the cellular membrane than short *b*PEI-2K chains for a given polymer weight concentration, especially when $C_{bPEI} \geq 2.7$ μg/mL, corresponding to N:P ≥ 10 in the typical gene transfection experiment. It was also found that long cationic *b*PEI-25K chains can reverse the charge of the synthetic phospholipid vesicles at a much lower concentration (~2 μg/mL) than their short counterparts [93]. It seems that the destabilization and/or disruption of the phospholipid membrane by long free PEI chains is correlated to the less entrapment of polyplexes into the acidic endolysosomes, and to some extent, the enhanced uptake of the polyplexes from the extracellular space into cells.

A sufficient amount of evidence has been accumulated to suggest that the escape of the polyplexes from endolysosomes by the osmotic-pressure-increase-induced membrane rupture might not be true and the buffer capacity of polycationic chains is only partially responsible for the safe trafficking of the polyplexes in the intracellular space. Recently, we proposed a hypothesis to account for why long polycationic chains free in the solution mixture are able to enhance the gene transfection [93]. Namely, instead of escaping from endolysosomes by virtue of the "proton sponge" effect, we think that long free polycationic chains embedded in or on the membranes might actually block the signal proteins for inter-vesicular fusion so that most of the endocytosis-ingested virginal vesicles with the polyplexes inside do not fuse with the endolysosomes by following the earlier-to-later endosome-endolysosome-lysosome pathway so that most of the polyplexes are not trapped and digested by enzymes inside lysosomes. This hypothesis will be addressed in details in the following chapters.

### 1.2.4 Nuclear Localization and Unloading of DNA

Once internalized into the cell and avoiding the entrapment inside endolysosomes, the polyplexes have to move towards and enter the cell nucleus and unload/release DNA inside for transcription. In the cytosolic transport step, the polyplexes could first escape from the endocytosis-ingested virginal vesicles or endosomes and then travel along microtubules, direct to the nucleus periphery, similar to adenoviruses [111], or first reach the nucleus periphery with the phospholipids coating and then release from these vesicles before entering the nucleus, resembling to adeno-associated viruses [112]. In either way, experimental evidences showed that the cytosolic delivery of polyplexes to the nucleus periphery is an active (not passive diffusive) process, with a linear speed of v $\sim$ $10^{-1}$ μm/s in both COS-7 [113] and HUH-7 cells [114], measured by a real-time multiple particle tracking (MPT) technique. It is worth-noting that normally such transportation is not a rate-limiting step in the intracellular trafficking. However, less attention has been paid to the subsequent nuclear localization, which imposes a great hurdle in the gene transfection [115]. The cell nucleus is separated from the cytoplasm by a double-layer membrane with tightly regulated pores that govern the import and export of a specific set of biomacromolecules (RNAs and proteins). The nuclear pore complexes (NPCs) allow passive diffusion of small molecules (diameter < 35 nm), while larger proteins have to be actively transported via specific nuclear proteins, such as importins [2]. Viruses have evolved functions to utilize this nuclear import machinery, but unmodified polymers or *p*DNA clearly have no such an ability. Early studies showed that the polyplexes (or *p*DNA) mainly entered the nucleus during the cell mitosis when the nuclear membrane temporarily dismantled [2, 116]. This partially explains why the gene transfection efficiency is extremely low when non-dividing or growth-arrested cells are used [116].

Many proteins are naturally targeted to the nucleus via some nuclear localization signals (NLS), short cationic peptides whose sequences are recognized by importins [2]. Using such a nuclear import machinery, one can attach a synthetic peptide with a NLS peptide to DNA so that the hybrid DNA-NLS can be identified as a nuclear import substrate [115]. Initial studies showed that the conjugation of a NLS peptide to a circular [117] or a linear DNA [118] enhanced the importin-induced nuclear translocation in the gene transfection. Recent studies also revealed that the nuclear factor kappa B (NF$\kappa$B), a family of transcription factors that shuttle between the cytoplasm and cell nucleus under specific conditions, is a desirable intracellular target to increase the nuclear import of *p*DNA [119]. The NF$\kappa$B binding sequences were optimized and constructed into *p*DNA, leading to an effective nuclear import and a prolonged in vivo transgene expression [120]. Note that in such a strategy, the release of *p*DNA from the polyplexes in the cytoplasm, preferably near the nuclear membrane, is a prerequisite. Jeong et al. [121] developed poly(amido ethylenimine), whose backbone is degradable in the cytoplasm by reduction, to facilitate the release of *p*DNA from the polyplexes in the cytoplasm. They showed that upon the activation of NF$\kappa$B by interleukin-1$\beta$, most of *p*DNA released due to the degradation of poly(amido ethylenimine) were translocated into the nucleus, leading to a much higher gene transfection efficiency in comparison with the PEI-mediated transfection. In another study, Choi et al. [122] improved the nuclear import of the polyplexes by attaching a glucocorticoid steroid molecule, dexamethasone, to *b*PEI-2K because dexamethasone can dilate the NPCs upon binding to its glucocorticoid receptor and thereby create a "giant pore" for impermeable macromolecules [123]. In this way, the dexamethasone-conjugated *b*PEI-2K and large *b*PEI-25K exhibited a similar gene transfection efficiency for higher N:P ratios but the *b*PEI-2K derivatives were much less cytotoxic.

Incorporating a viral component into a non-viral vector is another approach to enhance the nuclear translocation. Very recently, Pack et al. [124] constructed a hybrid polymer-virus vector by coating small non-infectious retroviral-like particles without a viral protein envelope with cationic PLL or PEI chains. The cationic polymer coatings are used to mediate the cellular uptake and the release of the hybrid particles from endosomes. Such hybrid vectors are efficient in the gene transfection, retain some important viral-like functions, including the nuclear import, genomic integration, and the infection of non-dividing cells; and at the same time, avoid some disadvantages inherent to the native viruses, such as the notorious fragility to physical forces in the common processing conditions.

In the extracellular space, we like to compact anionic DNA by cationic polymer chains as much as possible so that the polyplexes can be brought to cross the cell membrane and protected inside the cytoplasm before they hit the nuclear membrane, but then we wish the DNA/polymer complexation is weak so that DNA could be easily released for transcription inside the nucleus. Again, this is a narrow "slit" between these two requirements, another catch-22 problem [125, 126]. A quantitative comparison of the intracellular trafficking between adenovirus and non-viral vectors (cationic lipids and PEI) revealed that in addition to nuclear import, the rate-limiting step for non-viral gene delivery is the transcription and translation of exogenous DNA [95, 126], which might be related to the slow release of DNA from the polyplexes, i.e., the replacement of DNA by other polyanionic chains near or inside the nuclear membrane, presumably other proteins or DNA/RNA chains. This leads to another question: should DNA be released before or after its nuclear entry? Early studies saw that *p*DNA entered the cell nucleus together with its cationic vector [127], but later, it was found that *p*DNA was (at least partially) dissociated from the polyplexes upon their release from endosomes [8]. Using real-time CLSM, our recent studies revealed the existence of the released DNA in the cytoplasm as well as the polymer-bound DNA inside the nucleus. Nevertheless, most of DNA chains are still inside the polyplexes in the cytoplasm. Our studies also showed that the transgene expression was detectable as earlier as 6 h after the polyplexes administration and the corresponding transgene expression is $\sim$10 % of the maximum value at 36 h.

Currently, it is rather difficult, if not impossible, to elucidate whether this early transgene expression is prevailingly mediated by the *p*DNA released inside the cytoplasm or the nucleus, or both. We still question whether and how those polycationic chains free in the solution mixture play a vital role in this process. It is only generally known that for polymer-mediated gene transfection, the DNA payload should be properly "programmed" to release DNA after the polyplexes reach the nuclear membrane or inside the nucleus in order to enhance the gene transfection efficiency [8]. The advancements of modern analytic methods, such as live cell imaging with high spatio-temporal resolution, real-time particle tracking and intravital real-time CLSM, start to enable us to "see" the cytosolic and nuclear delivery of therapeutic genes, and more importantly, to elucidate how those free polycationic chains help the polyplexes to navigate through each of many "slits" in the intracellular space. Therefore, instead of synthesizing more polymeric vectors, it has come to the time for us to pay more attention on the detailed mechanism of the DNA releasing inside the intracellular space with well designed comparing/differentiating experiments so that future developments of non-viral polymer vectors can be better guided.

## 1.3 Objective and Main Achievements

The lack of safe, effective and controllable nucleic acid delivery vehicle remains a bottleneck to human gene therapy. Synthetic polymeric vectors, although safer than the viral carriers, generally do not possess the required efficacy, apparently due to a lack of functionality to overcome at least one of many intracellular gene-delivery obstacles. In this thesis, by using a combination of synthetic chemistry, polymer physics and cell biology, we design and execute decisive experiments to identify and address the following important questions with regard to the intracellular trafficking mechanism of DNA/polycation complexes; namely, (1) role of cationic chains free in the solution mixture of DNA and polymer in the gene transfection; (2) effect of length and topology of free cationic chains on endocytosis and the subsequent intracellular trafficking of polyplexes, particularly in the endolysosomal pathway; and (3) effect of endocytosis pathway on the intracellular fate of polyplexes.

The complexation and encapsulation of long anionic DNA with cationic polymer chains into small aggregates ($\sim 10^2$ nm) is the first and necessary step in the gene transfection. Our revisit of the complexation between DNA and polyethylenimine (PEI), one of the most efficient cationic polymeric vectors, by using a combination of laser light scattering and gel electrophoresis demonstrates that nearly all the DNA chains are complexed with PEI to form polyplexes when N:P $\sim$ 3, irrespective of the chain length of PEI and solvent used. However, a high in vitro gene transfection efficiency is only achieved when N:P $\geq$ 10. Putting these two facts together, it has been concluded that (1) each solution mixture with a higher N:P ratio actually contains two kinds of cationic chains: bound to DNA and free in the solution ($\sim$70 %); and (2) it is those free PEI chains that actually promote the gene transfection no matter whether they exist (are added) many hours before or after the administration of the polyplexes (N:P = 3).

Further, effects of the length and topology of both the bound and free polycationic chains on the gene transfection were respectively studied. Notably, both short ($\sim$2K) and long ($\sim$25K) PEI chains are capable of condensing DNA completely at N:P $\sim$ 3 but long free chains are $\sim 10^2$-fold more effective in enhancing the gene transfection, indicating that the length of free chains plays a vital role in the gene transfection. It is also interesting to note that for long free PEIs, the chain topology has nearly no effect on the transfection efficiency; but for short PEIs, linear free chains are ~10-fold more effective than branched ones. These results illustrate that bound PEI chains mainly play a role in DNA condensation and protection, and it is those polycationic chains free in the solution mixture that should get our attention.

To address how those free polymer chains facilitate the gene transfection, we quantitatively compared the cellular internalization and lysosomal distribution of polyplexes in the presence of different free PEIs by using a confocal image-assisted three-dimensionally integrated quantification (CIQ) method. Cellular uptake kinetics reveals that long free PEI chains boost the uptake rate, but their major contribution is in the intracellular space. In the endolysosomal pathway, the efficacy of free cationic chains is tightly correlated to their ability in preventing the entrapment of polyplexes into the acidic lysosomes. Namely, without free PEI chains or with short branched ones, the fraction of polyplexes inside lysosomes keeps escalating and reaches ~40 % in the first 6 h. In contrast, with short linear or long free chains, the percentage of polyplexes entrapped into lysosomes is reduced to just ~15 %. Notably, the popular "proton sponge" effect is not dominant in this process because the shut-down of proton pump only partially attenuates the transfection efficiency. It is postulated that free PEI chains with a proper length (1520 nm) are able to facilitate the intracellular trafficking via (1) blocking the signal proteins on the inner cell membrane so that the endolysosomes development is prevented or slowed down; and/or (2) promoting the escape from the virginal vesicles and early endosomes.

Finally, we explored the effect of endocytosis routes (clathrin and caveolae-mediated) on the intracellular trafficking of polyplexes. At N:P = 4, with limited amount of free PEIs, the transfection efficiency was almost completely lost when caveolae-mediated pathway was blocked, whereas the transfection efficiency was enhanced by ~3 times when clathrin-mediated (CME) pathway was inhibited. This indicates that at low N:P, most of polyplexes are entrapped into lysosomes in the CME pathway, and only the caveolae pathway leads to an effective gene transfection. On the other hand, at N:P = 7, these free PEIs are sufficient to prevent the entrapment of polyplexes into the later endolysosomes in the CME pathway, so the inhibition of caveolae pathway only reduced the gene transfection efficiency by ~2 times, and this decrease is possibly due to the reduced cellular uptake.

## References

1. Mulligan, R. (1993). *Science, 260*, 926–932.
2. Pack, D. W., Hoffman, A. S., Pun, S., & Stayton, P. S. (2005). *Nature Reviews Drug Discovery, 4*, 581–593.
3. Verma, I. M., & Somia, N. (1997). *Nature, 389*, 239–242.

4. Check, E. (2002). *Nature, 420*, 116–118.
5. Hacien-Bey-Abina, S. (2003). *Science, 302*, 568.
6. Raper, S. E., Chirmule, N., Lee, F. S., Wivel, N. A., Bagg, A., Gao, G. P., et al. (2003). *Molecular Genetics and Metabolism, 80*, 148–158.
7. Li, S. D., & Huang, L. (2007). *Journal of Controlled Release, 123*, 181–183.
8. Won, Y-. Y., Sharma, R., & Konieczny, S. F. (2009). *Journal of Controlled Release, 139*, 88–93.
9. Wu, G. Y., & Wu, C. H. (1987). *Journal of Biological Chemistry, 262*, 4429–4432.
10. Boussif, O., Lezoualch, F., Zanta, M. A., Mergny, M. D., Scherman, D., Demeneix, B., et al. (1995). *Proceedings of the National Academy of Sciences of the United States of America, 92*, 7297–7301.
11. Neu, M., Fischer, D., & Kissel, T. (2005). *Journal of Gene Medicine, 7*, 992–1009.
12. Pollard, H., Remy, J. S., Loussouarn, G., Demolombe, S., Behr, J. P., & Escande, D. (1998). *Journal of Biological Chemistry, 273*, 7507–7511.
13. Bieber, T., Meissner, W., Kostin, S., Niemann, A., & Elsasser, H. P. (2002). *Journal of Controlled Release, 82*, 441–454.
14. Oh, Y. K., Suh, D., Kim, J. M., Choi, H. G., Shin, K., & Ko, J. (2002). *Journal of Gene Therapy, 9*, 1627–1632.
15. Sonawane, N. D., Szoka, F. C., & Verkman, A. S. (2003). *Journal of Biological Chemistry, 278*, 44826–44831.
16. de Bruin, K. G., Fella, C., Ogris, M., Wagner, E., Ruthardt, N., & Brauchle, C. (2008). *Journal of Controlled Release, 130*, 175–182.
17. Wu, G. Y., & Wu, C. H. (1988). *Journal of Biological Chemistry, 263*, 14621–14624.
18. Haensler, J., & Szoka, F. C. (1993). *Bioconjugate Chemistry, 4*, 372–379.
19. Mintzer, M. A., & Simanek, E. E. (2009). *Chemical Reviews, 109*, 259–302.
20. Jones, G. D., Langsjoen, A., Neumann, M. M. C., & Zomlefer, J. (1944). *Journal of Organic Chemistry, 9*, 125–147.
21. Brissault, B., Kichler, A., Guis, C., Leborgne, C., Danos, O., & Cheradame, H. (2003). *Bioconjugate Chemistry, 14*, 581–587.
22. Klotz, I. M. (1968). Sloniews, Ar. *Biochemical and Biophysical Research Communications, 31*, 421–426.
23. Suh, J., Paik, H. J., & Hwang, B. K. (1994). *Bioorganic Chemistry, 22*, 318–327.
24. Behr, J. P. (1997). *Chimia, 51*, 34–36.
25. Fischer, D., Bieber, T., Li, Y. X., Elsasser, H. P., & Kissel, T. (1999). *Pharmaceutical Research, 16*, 1273–1279.
26. Godbey, W. T., Wu, K. K., & Mikos, A. G. (1999). Biomed. *Journal of Biomedical Materials Research, 45*, 268–275.
27. Reschel, T., Konak, C., Oupicky, D., Seymour, L. W., & Ulbrich, K. (2002). *Journal of Controlled Release, 81*, 201–217.
28. Kunath, K., von Harpe, A., Fischer, D., Peterson, H., Bickel, U., Voigt, K., et al. (2003). *Journal of Controlled Release, 89*, 113–125.
29. Moghimi, S. M., Symonds, P., Murray, J. C., Hunter, A. C., Debska, G., & Szewczyk, A. (2005). *Molecular Therapy, 11*, 990–995.
30. Parhamifar, L., Larsen, A. K., Hunter, A. C., Andresen, T. L., & Moghimi, S. M. (2010). *Soft Matter, 6*, 4001–4009.
31. Gao, X., Yao, L., Song, Q., Zhu, L., Xia, Z., Xia, H., et al. (2011). *Biomaterials, 32*, 8613–8625.
32. Gosselin, M. A., Guo, W. J., & Lee, R. J. (2001). *Bioconjugate Chemistry, 12*, 989–994.
33. Thomas, M., Ge, Q., Lu, J. J., Chen, J. Z., & Klibanov, A. M. (2005). *Pharmaceutical Research, 22*, 373–380.
34. Breunig, M., Lungwitz, U., Liebl, R., & Goepferich, A. (2007). *Proceedings of the National Academy of Sciences of the United States of America, 104*, 14454–14459.
35. Peng, Q., Zhong, Z. L., & Zhuo, R. X. (2008). *Bioconjugate Chemistry, 19*, 499–506.
36. Kang, H. C., Kang, H. J., & Bae, Y. H. (2011). *Biomaterials, 32*, 1193–1203.

37. Saito, G., Swanson, J. A., & Lee, K. D. (2003). *Advanced Drug Delivery Reviews, 55*, 199–215.
38. Lee, Y., Mo, H., Koo, H., Park, J. Y., Cho, M. Y., Jin, G. W., et al. (2007). *Bioconjugate Chemistry, 18*, 13–18.
39. Deng, R., Yue, Y., Jin, F., Chen, Y. C., Kung, H. F., Lin, M. C. M., et al. (2009). *Journal of Controlled Release, 140*, 40–46.
40. Wightman, L., Kircheis, R., Rossler, V., Carotta, S., Ruzicka, R., Kursa, M., et al. (2001). *Journal of Gene Medicine, 3*, 362–372.
41. Dash, P. R., Read, M. L., Barrett, L. B., Wolfert, M., & Seymour, L. W. (1999). *Gene Therapy, 6*, 643–650.
42. Ogris, M., Brunner, S., Schuller, S., Kircheis, R., & Wagner, E. (1999). *Gene Therapy, 6*, 595–605.
43. Petersen, H., Fechner, P. M., Martin, A. L., Kunath, K., Stolnik, S., Roberts, C. J., et al. (2002). *Bioconjugate Chemistry, 13*, 845–854.
44. Knorr, V., Allmendinger, L., Walker, G. F., Paintner, F. F., & Wagner, E. (2007). *Bioconjugate Chemistry; 18*, 1218–1225.
45. Blessing, T., Kursa, M., Holzhauser, R., Kircheis, R., & Wagner, E. (2001). *Bioconjugate Chemistry, 12*, 529–537.
46. Kursa, M., Walker, G. F., Roessler, V., Ogris, M., Roedl, W., Kircheis, R., et al. (2003). *Bioconjugate Chemistry, 14*, 222–231.
47. Zanta, M. A., Boussif, O., Adib, A., & Behr, J. P. (1997). *Bioconjugate Chemistry, 8*, 839–844.
48. Cheng, H., Zhu, J. L., Zeng, X., Jing, Y., Zhang, X. Z., & Zhuo, R. X. (2009). *Bioconjugate Chemistry, 20*, 481–487.
49. Gabrielson, N. P., & Pack, D. W. (2009). *Journal of Controlled Release, 136*, 54–61.
50. de Bruin, K., Ruthardt, N., von Gersdorff, K., Bausinger, R., Wagner, E., Ogris, M., et al. (2007). *Molecular Therapy, 15*, 1297–1305.
51. van de Wetering, P., Cherng, J. Y., Talsma, H., & Hennink, W. E. (1997). *Journal of Controlled Release, 49*, 59–69.
52. van de Wetering, P., Moret, E. E., Schuurmans-Nieuwenbroek, N. M. E., van Steenbergen, M. J., & Hennink, W. E. (1999). *Bioconjugate Chemistry, 10*, 589–597.
53. van de Wetering, P., Schuurmans-Nieuwenbroek, N. M. E., Hennink, W. E., & Storm, G. (1999). *Journal of Gene Medicine, 1*, 156–165.
54. Cai, J. G., Yue, Y. A., Rui, D., Zhang, Y. F., Liu, S. Y., & Wu, C. (2011). *Macromolecules, 44*, 2050–2057.
55. Layman, J. M., Ramirez, S. M., Green, M. D., & Long, T. E. (2009). *Biomacromolecules, 10*, 1244–1252.
56. Davis, M. E., & Brewster, M. E. (2004). *Nature Reviews Drug Discovery, 3*, 1023–1035.
57. Gonzalez, H., Hwang, S. J., & Davis, M. E. (1999). *Bioconjugate Chemistry, 10*, 1068–1074.
58. Davis, M. E. (2009). *Molecular Pharmaceutics, 6*, 659–668.
59. Pun, S. H., & Davis, M. E. (2002). *Bioconjugate Chemistry, 13*, 630–639.
60. Bellocq, N. C., Pun, S. H., Jensen, G. S., & Davis, M. E. (2003). *Bioconjugate Chemistry, 14*, 1122–1132.
61. Pun, S. H., Bellocq, N. C., Liu, A. J., Jensen, G., Machemer, T., Quijano, E., et al. (2004). *Bioconjugate Chemistry, 15*, 831–840.
62. Bartlett, D. W., & Davis, M. E. (2007). *Bioconjugate Chemistry, 18*, 456–468.
63. Bellocq, N. C., Davis, M. E., Engler, H., Jensen, G. S., Liu, A. J., Machemer, T., et al. (2003). *Molecular Therapy, 7*, S290–s290.
64. Pun, S. H., Tack, F., Bellocq, N. C., Cheng, J. J., Grubbs, B. H., Jensen, G. S., Davis, M. E., Brewster, M., Janicot, M., Janssens, B., Floren, W., & Bakker, A. (2004). *Cancer Biology and Therapy, 3*, 641–650.

65. Hu-Lieskovan, S., Heidel, J. D., Bartlett, D. W., Davis, M. E., & Triche, T. (2005). *Journal of Cancer Research, 65*, 8984–8992.
66. Bartlett, D. W., Su, H., Hildebrandt, I. J., Weber, W. A., & Davis, M. E. (2007). *Proceedings of the National Academy of Sciences of the United States of America, 104*, 15549–15554.
67. Heidel, J. D., Yu, Z. P., Liu, J. Y. C., Rele, S. M., Liang, Y. C., Zeidan, R. K., et al. (2007). *Proceedings of the National Academy of Sciences of the United States of America, 104*, 5715–5721.
68. Davis, M. E., Zuckerman, J. E., Choi, C. H. J., Seligson, D., Tolcher, A., Alabi, C. A., et al. (2010). *Nature, 464*, 1067–1071.
69. Grimm, D., & Kay, M. A. (2003). *Current Gene Therapy, 3*, 281–304.
70. Thomas, J. L., & Tirrell, D. A. (1992). *Accounts of Chemical Research, 25*, 336–342.
71. Thomas, J. L., Barton, S. W., & Tirrell, D. A. (1994). *Biophysical Journal, 67*, 1101–1106.
72. Murthy, N., Robichaud, J. R., Tirrell, D. A., Stayton, P. S., & Hoffman, A. S. (1999). *Journal of Controlled Release, 61*, 137–143.
73. Murthy, N., Chang, I., Stayton, P., & Hoffman, A. (2001). *Macromolecular Symposium, 172*, 49–55.
74. Cheung, C. Y., Murthy, N., Stayton, P. S., & Hoffman, A. S. (2001). *Bioconjugate Chemistry, 12*, 906–910.
75. Kiang, T., Bright, C., Cheung, C. Y., Stayton, P. S., Hoffman, A. S., & Leong, K. W. (2004). *Journal of Biomaterials Science, Polymer Edition, 15*, 1405–1421.
76. Kyriakides, T. R., Cheung, C. Y., Murthy, N., Bornstein, P., Stayton, P. S., & Hoffman, A. S. (2002). *Journal of Controlled Release, 78*, 295–303.
77. Convertine, A. J., Benoit, D. S. W., Duvall, C. L., Hoffman, A. S., & Stayton, P. S. (2009). *Journal of Controlled Release, 133*, 221–229.
78. Foster, S., Duvall, C. L., Crownover, E. F., Hoffman, A. S., & Stayton, P. S. (2010). *Bioconjugate Chemistry, 21*, 2205–2212.
79. Miyata, K., Oba, M., Nakanishi, M., Fukushima, S., Yamasaki, Y., Koyama, H., et al. (2008). *Journal of American Chemical Society, 130*, 16287–16294.
80. Itaka, K., Ishii, T., Hasegawa, Y., & Kataoka, K. (2010). *Biomaterials, 31*, 3707–3714.
81. Nomoto, T., Matsumoto, Y., Miyata, K., Oba, M., Fukushima, S., Nishiyama, N., et al. (2011). *Journal of Controlled Release, 151*, 104–109.
82. Richardson, S. C. W., Pattrick, N. G., Man, Y. K. S., Ferruti, P., & Duncan, R. (2001). *Biomacromolecules, 2*, 1023–1028.
83. Pattrick, N. G., Richardson, S. C. W., Casolaro, M., Ferruti, P., & Duncan, R. (2001). *Journal of Controlled Release, 77*, 225–232.
84. Wan, K. W., Malgesini, B., Verpilio, L., Ferruti, P., Griffiths, P. C., Paul, A., et al. (2004). *Biomacromolecules, 5*, 1102–1109.
85. Richardson, S. C. W., Pattrick, N. G., Lavignac, N., Ferruti, P., & Duncan, R. (2010). *Journal of Controlled Release, 142*, 78–88.
86. Bloomfield, V. A. (1997). *Biopolymers, 44*, 269–282.
87. Wagner, E., Cotten, M., Foisner, R., & Birnstiel, M. L. (1991). *Proceedings of the National Academy of Sciences of the United States of America, 88*, 4255–4259.
88. Clamme, J. P., Azoulay, J., & Mely, Y. (2003). *Biophysical Journal, 84*, 1960–1968.
89. Clamme, J. P., Krishnamoorthy, G., & Mely, Y. (2003). *Biochimica et Biophysica Acta (BBA), Biomembranes, 1617*, 52–61.
90. Boeckle, S., von Gersdorff, K., van der Piepen, S., Culmsee, C., Wagner, E., & Ogris, M. (2004). *Journal of Gene Medicine, 6*, 1102–1111.
91. Fahrmeir, J., Gunther, M., Tietze, N., Wagner, E., & Ogris, M. (2007). *Journal of Controlled Release, 122*, 236–245.
92. Yue, Y., Jin, F., Deng, R., Cai, J., Chen, Y., Lin, M. C. M., et al. (2011). *Journal of Controlled Release, 155*, 67–76.

93. Yue, Y. A., Jin, F., Deng, R., Cai, J. G., Dai, Z. J., Lin, M. C. M., et al. (2011). *Journal of Controlled Release, 152*, 143–151.
94. Khalil, I. A., Kogure, K., Akita, H., & Harashima, H. (2006). *Pharmacological Review, 58*, 32–45.
95. Goncalves, C., Mennesson, E., Fuchs, R., Gorvel, J. P., Midoux, P., & Pichon, C. (2004). *Molecular Therapy, 10*, 373–385.
96. Bishop, N. E. (1997). *Reviews in Medical Virology, 7*, 199–209.
97. Pelkmans, L., & Helenius, A. (2002). *Traffic, 3*, 311–320.
98. Conner, S. D., & Schmid, S. L. (2003). *Nature, 422*, 37–44.
99. Medina-Kauwe, L. K., Xie, J., & Hamm-Alvarez, S. (2005). *Gene Therapy, 12*, 1734–1751.
100. Parton, R. G., & Simons, K. (2007). *Nature Reviews Molecular Cell Biology, 8*, 185–194.
101. Pelkmans, L., Burli, T., Zerial, M., & Helenius, A. (2004). *Cell, 118*, 767–780.
102. Joiner, K. A., Fuhrman, S. A., Miettinen, H. M., Kasper, L. H., & Mellman, I. (1990). *Science, 249*, 641–646.
103. Rejman, J., Bragonzi, A., & Conese, M. (2005). *Molecular Therapy, 12*, 468–474.
104. van der Aa, M. A. E. M., Huth, U. S., Hafele, S. Y., Schubert, R., Oosting, R. S., Mastrobattista, E., et al. (2007). *Pharmaceutical Research, 24*, 1590–1598.
105. Thomas, M., & Klibanov, A. M. (2002). *Proceedings of the National Academy of Sciences of the United States of America, 99*, 14640–14645.
106. Forrest, M. L., Meister, G. E., Koerber, J. T., & Pack, D. W. (2004). *Pharmaceutical Research, 21*, 365–371.
107. Funhoff, A. M., van Nostrum, C. F., Koning, G. A., Schuurmans-Nieuwenbroek, N. M. E., Crommelin, D. J. A., & Hennink, W. E. (2004). *Biomacromolecules, 5*, 32–39.
108. Dubruel, P., Christiaens, B., Vanloo, B., Bracke, K., Rosseneu, M., Vandekerckhove, J., et al. (2003). *European Journal of Pharmaceutical Sciences, 18*, 211–220.
109. Dubruel, P., Christiaens, B., Rosseneu, M., Vandekerckhove, J., Grooten, J., Goossens, V., et al. (2004). *Biomacromolecules, 5*, 379–388.
110. Needham, D., & Nunn, R. S. (1990). *Biophysical Journal, 58*, 997–1009.
111. Leopold, P. L., Kreitzer, G., Miyazawa, N., Rempel, S., Pfister, K. K., Rodriguez-Boulan, E., et al. (2000). *Human Gene Therapy, 11*, 151–165.
112. Seisenberger, G., Ried, M. U., Endress, T., Buning, H., Hallek, M., & Brauchle, C. (2001). *Science, 294*, 1929–1932.
113. Suh, J., Wirtz, D., & Hanes, J. (2003). *Proceedings of the National Academy of Sciences of the United States of America, 100*, 3878–3882.
114. Bausinger, R., von Gersdorff, K., Braeckmans, K., Ogris, M., Wagner, E., Brauchle, C., et al. (2006). *Angewandte Chemie International Edition, 45*, 1568–1572.
115. Dean, D. A., Strong, D. D., & Zimmer, W. E. (2005). *Gene Therapy, 12*, 881–890.
116. Grosse, S., Thevenot, G., Monsigny, M., & Fajac, I. (2006). *Journal of Gene Medicine, 8*, 845–851.
117. Branden, L. J., Mohamed, A. J., & Smith, C. I. E. (1999). *Nature Biotechnology, 17*, 784–787.
118. Zanta, M. A., Belguise-Valladier, P., & Behr, J. P. (1999). *Proceedings of the National Academy of Sciences of the United States of America, 96*, 91–96.
119. Choi, S., Oh, S., Lee, M., & Kim, S. W. (2005). *Molecular Therapy, 12*, 885–891.
120. Goncalves, C., Ardourel, M. Y., Decoville, M., Breuzard, G., Midoux, P., Hartmann, B., et al. (2009). *Journal of Gene Medicine, 11*, 401–411.
121. Jeong, J. H., Kim, S. H., Christensen, L. V., Feijen, J., & Kim, S. W. (2010). *Bioconjugate Chemistry, 21*, 296–301.
122. Bae, Y. M., Choi, H., Lee, S., Kang, S. H., Kim, Y. T., Nam, K., et al. (2007). *Bioconjugate Chemistry, 18*, 2029–2036.
123. Kastrup, L., Oberleithner, H., Ludwig, Y., Schafer, C., & Shahin, V. (2006). *Journal of Cell Physiology, 206*, 428–434.

124. Ramsey, J. D., Vu, H. N., & Pack, D. W. (2010). *Journal of Controlled Release, 144*, 39–45.
125. Honore, I., Grosse, S., Frison, N., Favatier, F., Monsigny, M., & Fajac, I. (2005). *Journal of Controlled Release, 107*, 537–546.
126. Cohen, R. N., van der Aa, M. A. E. M., Macaraeg, N., Lee, A. P., & Szoka, F. C. (2009). *Journal of Controlled Release, 138*, 185.
127. Godbey, W. T., Wu, K. K., & Mikos, A. G. (1999). *Proceedings of the National Academy of Sciences of the United States of America, 96*, 5177–5181.

# Chapter 2
# Revisiting Complexation Between DNA and Polyethylenimine: The Effect of Uncomplexed Chains Free in the Solution Mixture on Gene Transfection

## 2.1 Introduction

In comparison with viral vectors, more efforts have recently been spent on the development of non-viral vectors because of few fatal accidents in clinical trials of viral carriers [1–3]. It has been well recognized that non-viral vectors have their own advantages, such as low immune toxicity, construction flexibility and facile fabrication, in the gene transfection, especially for clinical applications [4–6]. However, they are still much less efficient than their viral counterparts. Among thousands of experimentally tested non-viral vectors, PEI is still considered as one of the most efficient candidates to deliver genes and often served as a "golden standard" [7–9]. Previously, PEI has been chemically modified in different ways so that additional functions were introduced for a better gene delivery, including the incorporation of intracellular biodegradable linkers [10, 11], the PEGylation to improve the serum stability during circulation [12, 13] and the attachment of some functional molecules to target specific cells or tissues [12–15]. Less attention, however, has been paid to why PEI remains one of the best non-viral carriers and how it facilitates the intracellular trafficking [16–27].

The first step in the development of non-viral vectors is how to package long anionic DNA chains into a small particle, i.e., the DNA complexation and condensation. Previous study showed that the physical and colloidal characteristics of resultant polymer/DNA polyplexes are important for an effective delivery of gene [28–32], but results were controversial and not conclusive. Much of the past effort has been devoted to the synthesis of state-of-the-art polymers [33–37] and subsequent polyplex formation [38–41], rather than the fundamental understanding of how a non-viral vector promotes the intracellular trafficking of DNA besides its roles in DNA encapsulation and protection. The proposed proton-sponge effect has been well accepted and taken as granted by those who entered this field later [7, 42].

Using a combination of different methods to characterize the size, molar mass and surface charge of the DNA/PEI polyplexes formed under different conditions, we confirmed some of the previous literature results, but revealed that the average size and chain density of the resultant polyplexes are not that important in the gene

Y. Yanan, *How Free Cationic Polymer Chains Promote Gene Transfection*, Springer Theses, DOI: 10.1007/978-3-319-00336-8_2,

transfection. In order to achieve an optimal transfection efficiency, the polyplexes in the solution mixture should have a slightly positively charged surface with a N:P ratio higher than 10. It is understandable that a positively charged surface facilitates the attachment of polyplexes to the negatively charged cellular membrane so that the endocytosis is promoted. However, it is still not clear why a high N:P ratio is required. Few previous studies showed that when the N:P ratio is high, a large amount of PEI chains are free in the solution mixture of DNA and PEI [43, 44]. It was also revealed that these excessive polycationic chains are much more toxic than those bound to DNA inside the polyplexes [44, 45]. Moreover, Wagner et al. found that the removal of free PEI chains by size exclusion chromatography significantly reduced the gene transfection efficiency [44]. To our knowledge, only few previous studies have noted such an effect of free polycationic chains on the gene transfection [43–47]. Most of the researchers in this field have overlooked such a finding. The current study was designed to elucidate the impact of polymer chains free in the solution mixture on gene delivery.

## 2.2 Materials and Methods

### *2.2.1 Materials and Cell Lines*

Branched PEI with a weight-averaged molar mass of 25,000 g/mol (*b*PEI-25 K, Sigma-Aldrich) was used without further purification. Initial plasmid DNA pGL3-control vector (5,256 bp) encoding modified firefly luciferase and pEGFP-N1 (4,700 bp) expressing enhanced green fluorescent protein (EGFP) were purchased from Promega (USA) and Clontech (Germany), respectively. A large amount of these plasmids were prepared using a Qiagen Plasmid Maxi Kit (Qiagen, Germany). POPO-3 iodide, fluorescein-5-isothiocyanate (FITC) and di-4-ANEPPDHQ were purchased from Invitrogen (USA). Fetal bovine serum (FBS), Dulbecco's modified Eagle's medium (DMEM) and penicillin–streptomycin were purchased from GIBCO (USA). 293T cells were grown at 37 °C, 5 % $CO_2$ in DMEM supplemented with 10 % FBS, penicillin at 100 units/mL and streptomycin at 100 μg/mL.

### *2.2.2 Formation of DNA/PEI Polyplexes*

Plasmid DNA was complexed with PEI in either distilled water or phosphate buffered saline (PBS) to form the DNA/PEI polyplexes as follows. Different amounts of PEI solution ($C = 1 \times 10^{-4}$–$1 \times 10^{-3}$ g/mL) were added dropwise into a dilute DNA solution ($C$ = 14.5 μg/mL), resulting in different molar ratios of nitrogen from PEI to phosphate from DNA (N:P). Each resultant DNA/PEI dispersion was incubated for 5 min at room temperature before its administration to the cell culture medium.

### 2.2.3 DNA Binding Assay

The binding of DNA to PEI to form polyplexes in the solution mixture was evaluated by the gel-shift assay. Each DNA/PEI dispersion with a desired N:P ratio was mixed with 6 × loading buffer and then loaded on a 0.8 % (w/v) agarose gel containing ethidium bromide in tris–borate EDTA buffer. The amount of DNA loaded into each well was 0.4 μg in a total volume of 10 μL. The electrophoresis was performed under 100 V for 45 min. DNA bands were visualized under UV. The DNA binding was also characterized using laser light scattering (LLS).

### 2.2.4 POPO-3 Fluorescence Quenching Assay

Plasmid was labeled with POPO-3 dye at 1 dye to 100 base pair ratio. DNA/PEI polyplexes with different N:P ratios were prepared as described above. The final DNA concentration is 1.45 μg/100 μL solution. After 5-min incubation, the relative fluorescence intensity ($F$) of each solution mixture was collected using a Hitachi F4500 fluorescence spectrophotometer (excitation 530 nm, emission 570 nm). The fraction of the uncomplexed DNA chains ($x$) was determined by $x = (F_{\text{polyplex}}-F_{\text{blank}})/(F_{\text{DNA}}-F_{\text{blank}})$.

### 2.2.5 Laser Light Scattering

A commercial LLS instrument (ALV5000) with a vertically polarized 22-mV He–Ne laser (632.8 nm, Uniphase) was used. The measurable angular range is 15–154°. In dynamic LLS, the intensity–intensity time correlation function ($G^{(2)}(\tau)$) of each DNA/PEI polyplex solution mixture was measured at different scattering angles. The Laplace inversion of each $G^{(2)}(\tau)$ can lead to a line-width distribution of $G(\Gamma)$ by the CONTIN program or by the double-exponential fitting, if there are only two relaxation modes, as

$$\{[G^{(2)}(q,\tau) - B]/B\}^{1/2} = A_1(q)e^{-<\Gamma>_1\tau} + A_2(q)e^{-\langle\Gamma\rangle_2\tau} \tag{2.1}$$

where $B$ is the measured baseline; $<\Gamma>$ and $A(q)$ are the average line-width and the normalized intensity contribution of each relaxation mode, respectively, and $A_1(q) + A_2(q) \equiv 1$; $q$ is the scattering vector defined as $q \equiv (4\pi n/\lambda_0)\sin(\theta/2)$ with $\theta$, $\lambda_0$ and $n$, the scattering angle, the incident wavelength in vacuum and the refractive index of solvent, respectively. $<\Gamma>$ can be related to the average translational diffusion coefficient $<D>$ as $<\Gamma> = <D> q^2$. Using the Stokes–Einstein equation [48], $<D>$ is further related to the average hydrodynamic radius ($<R_h>$) by $<R_h> = k_BT/(6\pi\eta <D>)$, where $\eta$ is the solvent viscosity. In this way, each $G(\Gamma)$ can also be converted into a hydrodynamic radius distribution $f(R_h)$.

On the other hand, the time-averaged scattering intensity from each relaxation mode can be calculated from the total time-averaged scattering intensity ($<I(q)>$) and $A(q)$, respectively, measured from static and dynamic LLS, i.e., $<I(q)>_1 = <I(q)>A_1(q)$ and $<I(q)>_2 = <I(q)>A_2(q)$. The plot of $1/<I(q)>_1$ or $1/<I(q)>_2$ versus $q^2$ leads to an average radius of gyration ($<R_g>_1$ or $<R_g>_2$) and the time-averaged scattering intensity at $q \rightarrow 0$ ($<I(0)>_1$ or $<I(0)>_2$) using [48, 49]

$$\frac{1}{<I(q)>} \approx \frac{1}{<I(0)>}\left(1 + \frac{1}{3} <R_g^2> q^2\right) \tag{2.2}$$

Where $<I(0)>$ is proportional to the weight-averaged molar mass. Therefore, individual short PEI chains with a much lower molar mass are invisible in LLS when long DNA chains exist in the solution mixture. The formation of DNA/PEI polyplexes with some DNA chains collapsed inside leads to a relaxation mode faster than that of individual swollen DNA chains. This is why we mainly focused on how N:P ratio affects the fast mode in the current study.

### 2.2.6 *Zeta-Potential Measurement*

The average mobility ($\mu_E$) of the polyplexes under an electric field in an aqueous solution was determined from the frequency shift in a laser Doppler spectrum using a commercial zeta-potential spectrometer (ZetaPlus, Brookhaven) with two platinum-coated electrodes and one He–Ne laser as the light source. Each data point in the mobility measurement was averaged over 20 times at 20 °C. The zeta-potential ($\zeta_{potential}$) can be calculated from $\mu_E$ using $\mu_E = 2\varepsilon\zeta_{potential} f(\kappa R_b)/(3\eta)$, where $\varepsilon$ is the permittivity of water and $1/\kappa$ is the Debye screening length [50]. When $\kappa R_b << 1$ (the Hückel limit), $f(\kappa R_b) \approx 1$; while $\kappa R_b >> 1$ (the Smoluchowski limit), $f(\kappa R_b) \approx 1.5$. In the current study, the Hückel and Smoluchowski limits are respectively used to calculate $\zeta_{potential}$ in water and PBS.

### 2.2.7 *Determination of the Content of Free PEI in DNA/PEI Dispersion*

The content of free PEI in the DNA/PEI dispersion was determined using a combination of filtration and copper complex assay. The DNA/PEI dispersion at N:P = 10 was prepared as described above, with a final concentration of 219 μg pGL3/1 mL PBS. After 4-h incubation, the DNA/PEI dispersion was centrifuged and the supernatant was carefully filtered using a 20-nm filter. The content of uncomplexed PEI chains in the filtrate was determined using a copper complex assay by adding 100 μL of the filtrate into 900 μL of 0.02 M cupric acetate solution containing 5 % potassium acetate (pH 5.5). The absorption of the filtrate was recorded at 630 nm ($A_{630}$) using a UV–Vis spectrometer (Hitachi, Japan), and converted to the corresponding PEI concentration according to the calibration curve.

### 2.2.8 *In Vitro Gene Transfection*

The in vitro gene transfection efficiency was quantified by using the luciferase transfection assays, in which plasmid pGL3 was used as an exogenous reporter gene. 293T cells were plated in 48-well plate at an initial density of $1.5 \times 10^5$ per well, 24 h prior to transfection. The DNA/PEI dispersion with a desired N:P ratio was further diluted in serum-free medium and then added at a final concentration of 0.4 μg DNA per well. The complete DMEM medium (800 μL per well) was added 6 h after the gene transfection. Using a GloMax 96 microplate luminometer (Promega, USA) and the Bio-Rad protein assay reagent, we respectively determined the transgene expression level and corresponding protein concentration in each well 48 h after the polyplex administration. The gene transfection efficiency is expressed as a relative luminescence unit (RLU) per cellular protein (mean ± SD of triplicates). Alternatively, the transfection efficiency can be directly visualized under a fluorescence microscopy when the plasmid pEGFP-N1 is used.

### 2.2.9 *Cytotoxicity Assay*

The cytotoxicity of free PEI chains and DNA/PEI dispersion was evaluated on 293T cells by using the MTT assay. 293T cells were seeded in a 96-well plate at an initial density of 5000 cells/well. After 24 h, free PEI chains alone or with DNA to form DNA/PEI dispersion were respectively added to cells at different chosen concentrations and N:P ratios. For the DNA/PEI dispersion, the final DNA concentration is 0.2 μg/well in a total volume of 100 μL. The treated cells were incubated in a humidified environment with 5 % $CO_2$ at 37 °C for 48 h. The MTT reagent (in 20 μL PBS, 5 mg/mL) was then added to each well. The cells were further incubated for 4 h at 37 °C. The medium in each well was then removed and replaced by 100 μL DMSO. The plate was gently agitated for 15 min before the absorbance ($A$) at 490 nm was recorded by a microplate reader (Bio-rad, USA). The cell viability ($y$) was calculated by $y = (A_{treated}/A_{control}) \times 100\ \%$, where $A_{treated}$ and $A_{control}$ are the absorbance of the cells cultured with polymer/polyplex and fresh culture medium, respectively. Each experiment condition was done in quadruple. The data was shown as the mean value plus a standard deviation (±SD).

### 2.2.10 DNA, PEI Labeling and Flow Cytometry

Plasmid pGL3 was covalently labeled with the fluorophore Cy5 using the Label IT kit (Mirus, Madison, WI) according to the manufacturer's instructions. For labeling of PEI, 11.2 mg of FITC (28.8 μmol) was mixed with 360 mg of *b*PEI-25 K (14.4 μmol) in EtOH, and the reaction mixture was stirred overnight at room temperature in the dark place. In order to separate labeled PEI from unreacted FITC, size exclusion chromatography (SEC) was performed using a gel-filtration column (Sephadex G-25 superfine) preequilibrated with HBS (20 mM HEPES, pH 7.4, 150 mM NaCl). The separated sample was freeze-dried and resolved in water. The gel electrophoresis results confirmed that the labeling on DNA and PEI chains has no obvious impact on the formation of DNA/PEI polyplexes (data not shown).

To explore the effect of free PEI chains on the cellular uptake of polyplexes, FITC-PEI, Cy5-pGL3/PEI polyplexes alone (N:P = 3) or polyplexes plus 7 portions of free FITC-PEI (N:P = 10) were added to the cells in serum-free DMEM and, after incubation at 37 °C, cells were harvested at the indicated time points. Briefly, cells were first rinsed twice with PBS containing 0.001 % SDS and then PBS to remove the extracellularly attached polyplexes [51]. Further, cells were detached by 0.05 % trypsin/EDTA supplemented with 20 mM sodium azide to prevent further endocytosis [52]. Finally, cells were washed twice by pelleting and then resuspended in ice-cold PBS containing 2 % FBS. Confocal microscopy images of cells were taken to ensure the effective removal of polyplexes from the cell surfaces (data not shown). Cellular uptake of polyplexes and free PEI chains was assayed by flow cytometry using a FC 500 flow cytometry system (Beckman Coulter, USA). The fluorophores FITC and Cy5 were excited at 488 and 635 nm, respectively, and the corresponding emissions were detected at 525/10 and 675/15 nm, respectively. To discriminate viable cells from dead cells and to exclude doublets, the cells were appropriately gated by forward/side scattering and pulse width. $1 \times 10^4$ gated events per sample were collected. Experiments were performed at least in triplicates.

### 2.2.11 Confocal Laser Scanning Microscopy

$6 \times 10^5$ cells were seeded in a $\mu$-Dish$^{35mm,\ high}$ (ibidi GmbH, Germany). After 24 h, the cellular membrane was stained with di-4-ANEPPDHQ for 15 min and then washed 2–3 times. Afterwards, the cell culture medium was carefully aspirated and FITC-labeled *b*PEI-25 K in serum-free DMEM was applied at a final concentration of $1 \times 10^{-5}$ g/mL. Live cell imaging was performed for 1.5 h using a Nikon C1si confocal laser scanning microscope equipped with a spectral imaging detector (Nikon, Japan) and a INU stage-top incubator (Tokai Hit, Japan). Image sequences were captured at approximately 30 s intervals. FITC and di-4-ANE-PPDHQ were visualized by the 488-nm excitation and the corresponding emission spectral was acquired in the 495–650 nm range at 5-nm wavelength resolution. The mean FITC-PEI fluorescence intensities at the cellular membrane and inside the cell are recorded at 520 nm and normalized by that outside the cell. All the spectral data was analyzed using the Nikon EZ-C1 software.

## 2.3 Results and Discussion

To investigate the complexation profiles of PEI-mediated vectors, we first monitored the formation of DNA/PEI polyplexes in PBS using a combination of static and dynamic LLS. Figure 2.1 shows the N:P ratio dependence of hydrodynamic radius distribution ($f(R_h)$) of pGL3 without and with the addition of different amounts of PEI. In the pure pGL3 solution (N:P = 0.00), the peak located at ~1.4 μm represents the swollen and extended DNA chains. The addition of a small amount of branched PEI (N:P = 0.25) leads to a new peak located at ~120 nm, presumably corresponding to the newly formed contracted DNA/PEI polyplexes. The shifting of the DNA peak from ~1.4 to ~1 μm indicates the partial contraction of DNA chains due to the incomplete complexation. Further addition of PEI leads to polyplexes of a larger size with more DNA and PEI chains incorporated into the individual complexes. When N:P ~ 3, the DNA peak completely disappears, suggesting that nearly all the DNA chains are complexed with PEI, i.e., no DNA chains are free in the solution mixture.

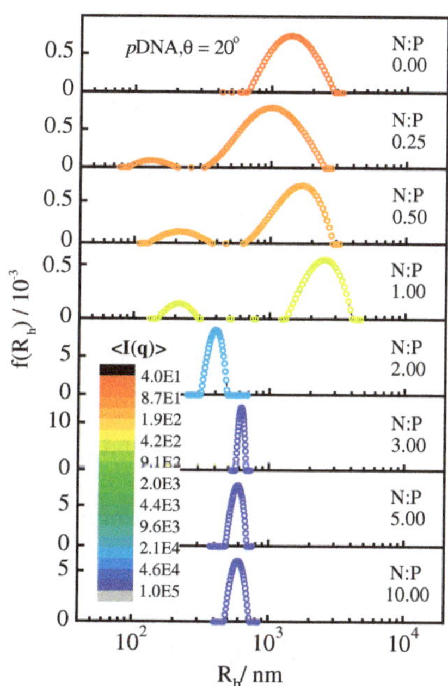


**Fig. 2.1** N:P ratio dependence of the normalized hydrodynamic radius distribution ($f(R_h)$) of DNA/PEI polyplexes formed in PBS, where different colors represent variation of the time-averaged scattering intensity of the solution mixture ($<I(q)>$)

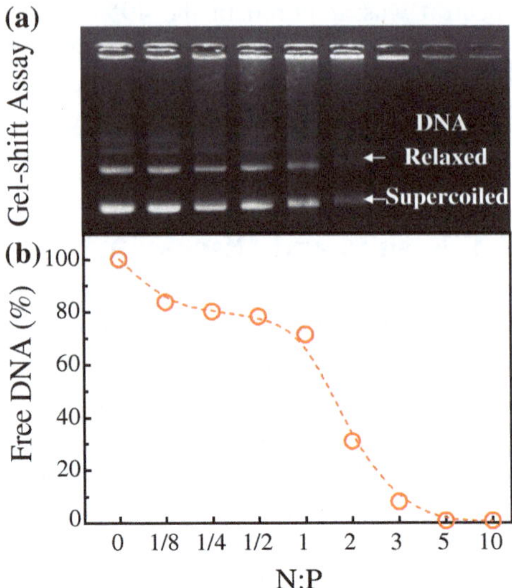


**Fig. 2.2** Complexation profile between PEI and DNA in PBS at different N:P ratios, evaluated by **a** gel-shift assay and **b** POPO-3 fluorescence quenching assay

The complexation between PEI and DNA was also evaluated by the gel-shift assay (Fig. 2.2a). Progressive condensation can be revealed from the retarded mobility of the DNA bands and their reduced fluorescence intensity. Clearly, the two DNA bands (supercoiled and relaxed forms) disappear when N:P $\sim$ 3. Furthermore, we quantitatively estimated the extent of DNA complexation at each given N:P ratio using a double-strand DNA intercalating dye, POPO-3, whose fluorescence is dramatically increased upon its binding to DNA. The complexation of DNA with PEI excludes POPO-3 molecules from the DNA double helix, resulting in a reduced fluorescence emission intensity. Figure 2.2b shows that over 90 % of the DNA chains have been condensed into the polyplexes when N:P $\sim$ 3. In static LLS, we also found that the average scattering intensity ($<I(0)>$) nearly remains a constant after N:P $\geq$ 3.

It is known that $<I(0)> \sim KCM_w \sim (dn/dc)^2CM_w$. Therefore, we can calculate the apparent weight-averaged molar mass ($M_{w,\ polyplex}$) of the polyplexes at each given N:P ratio using

$$M_{w,polyplex} = M_{w,DNA} \times \frac{<I(0)>_{polyplex} C_{DNA} (\frac{dn}{dc})^2_{DNA}}{<I(0)>_{DNA} C_{polyplex} (\frac{dn}{dc})^2_{polyplex}} \tag{2.3}$$

where $C$ is concentration and $(dn/dc)$ is the differential refractive index increment. Note that $(dn/dc)_{polyplex} = w_{DNA}(dn/dc)_{DNA} + w_{PEI}(dn/dc)_{PEI}$, where $w_{DNA}$ and $w_{PEI}$ are two normalized weight fractions of DNA and PEI inside the polyplexes, i.e., $w_{DNA} + w_{PEI} = 1$. It is necessary to estimate $C_{polyplex}$ and $w_{PEI}$ in order to calculate $M_{w,\ polyplex}$. When N:P $\leq$ 3, we can assume that nearly all the PEI chains are complexed with DNA, while the fraction of DNA associated with PEI can be estimated using the POPO-3 fluorescence assay mentioned above. Therefore,

$$C_{\text{polyplex}} - C_{\text{PEI}} + (1 - x)C_{\text{DNA}} \text{ and } W_{\text{PEI}} = C_{\text{PEI}}/C_{\text{polyplex}} \tag{2.4}$$

where $C_{\text{PEI}}$ and $C_{\text{DNA}}$ are the initial PEI and DNA concentrations, respectively; and $x$ is the fraction of those uncomplexed DNA chains in the solution mixture. On the other hand, when $3 < \text{N:P} \leq 10$, nearly all the DNA chains are complexed with PEI so that $C_{\text{polyplex}}$ is a constant in this range and $C_{\text{polyplex}} \approx C_{\text{polyplex}}$ (N:P = 3).

Figure 2.3 summaries the N:P ratio dependence of the average hydrodynamic radius ($\langle R_{\text{h}} \rangle$), the apparent weight-averaged molar mass ($M_{\text{w, polyplex}}$), the zeta-potential ($\zeta_{\text{potential}}$) and the gene transfection efficiency (RLU/mg protein) of the DNA/PEI polyplexes formed in PBS. The initial decrease of $\langle R_{\text{h}} \rangle$ and increases of both $M_{\text{w, polyplex}}$ and $\zeta_{\text{potential}}$ upon the addition of PEI (N:P = 0.125–1.50) indicate that the positively charged PEI chains are gradually complexed with the negatively charged DNA chains so that both PEI and DNA contract in the solution mixture. The sharp increases of both $\langle R_{\text{h}} \rangle$ and $M_{\text{w, polyplex}}$ at N:P ~ 2 reflect the merge of initially formed small polyplexes. Further addition of PEI in the range of N:P > 3 has nearly no effect on $\langle R_{\text{h}} \rangle$, $M_{\text{w, polyplex}}$ and $\zeta_{\text{potential}}$, clearing revealing that those PEI chains added afterwards are free in the solution mixture.

Only few previous studies have noted the existence of free PEI, and estimated the amount of those free polycationic chains in the DNA/PEI dispersion used for gene delivery. Clamme et al. found that, by using a two photon fluorescence correlation spectroscopy, ~86 % of the PEI chains are in the free form at N:P = 10 (i.e., polyplexes is completely formed at N:P ~ 1.4) [43], whereas Wagner and his coworkers showed that the purified polyplexes by SEC have an ultimate N:P ratio of ~2.5 [44]. Our LLS results here clearly shows that the anionic DNA chains are not fully condensed by the cationic PEI chains when N:P ~ 1.4, also reflected in its corresponding negative zeta-potential of the resultant polyplexes. Besides LLS, herein we used a simple filtration method combined with copper complex assay to determine the content of uncomplexed PEI chains in the DNA/PEI dispersion. Figure 2.4 shows that the PEI concentration after the removal of the polyplexes by a 20-nm filter is ~68.6 % of the initial PEI concentration at N:P = 10, further confirming that DNA is fully condensed by PEI only when N:P ≥ 3. For N:P = 10, ~70 % of the PEI chains are free in the solution mixture of PEI and DNA. On the other hand, our results in Fig. 2.3d consolidate that the polyplexes with 7 portions of free PEI chains (N:P = 10) are ~$10^3$ times more efficient in the gene transfection than those formed at N:P = 3 without free PEI chains.

Previous literature also repeatedly showed that when N:P ≥ 3, nearly all the DNA chains are completely complexed with PEI, but a higher N:P ratio leads to much better gene transfection. Putting these two experimental facts together, one should ask an obvious, but certainly overlooked, question; namely, is it the free PEI chains, rather than those inside the DNA/PEI polyplexes, that play a vital role in the gene transfection? To generalize such a question, we have tested PEI-mediated vectors with different chain lengths in both salt-free water and salt-rich PBS. All of our results (not shown here) confirm that the solution mixture with polyplexes alone have a much lower gene transfection efficiency.

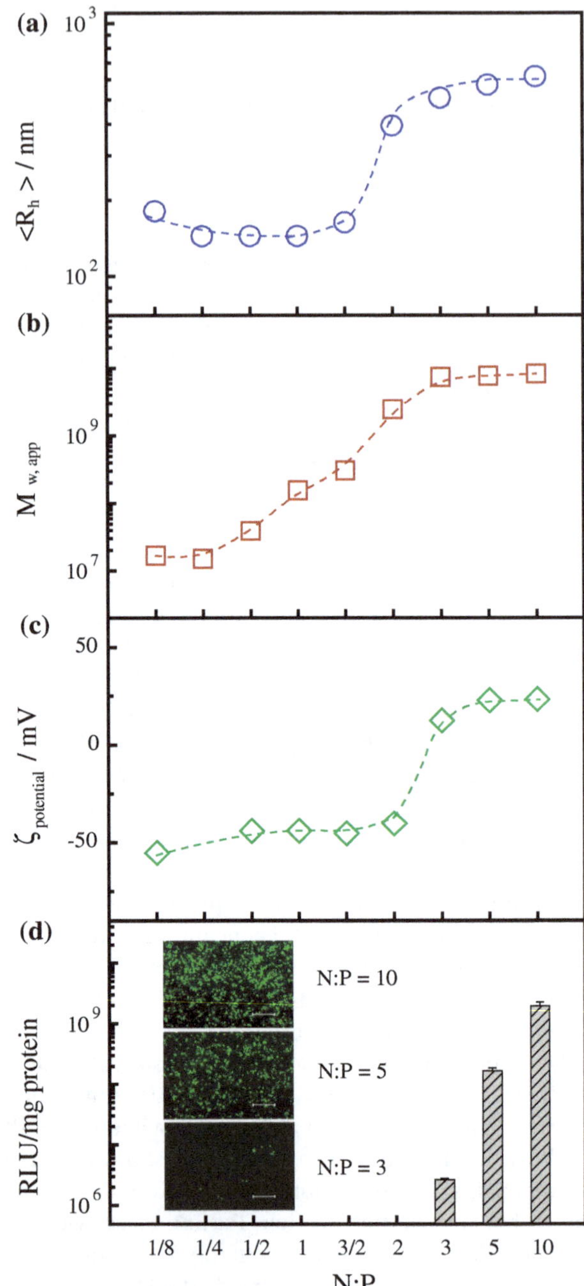


**Fig. 2.3** N:P ratio dependence of **a** average hydrodynamic radius ($<R_h>$); **b** apparent weight-averaged molar mass ($M_{w,\ polyplex}$); **c** average zeta-potential ($\zeta_{potential}$); and **d** gene transfection efficiency (RLU/mg protein) of polyplexes formed in PBS. *Inset* Fluorescent microscopic image of 293T cells transfected with pEGFP-N1 after 48 h. Scale bar: 200 μm

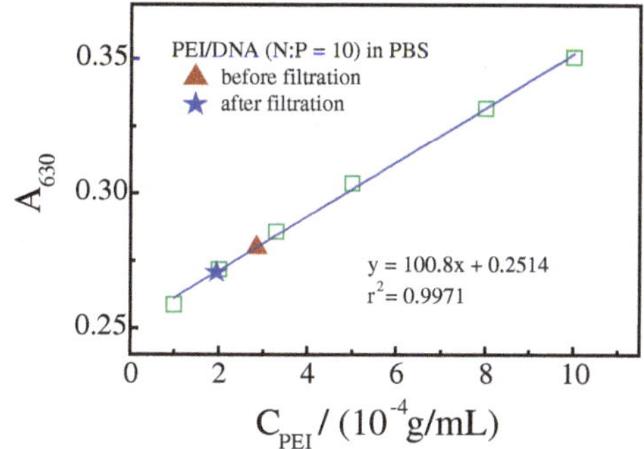


**Fig. 2.4** Concentration calibration curve of *b*PEI-25 K chains in PBS, where symbols ▲ and ★ refer to the concentration of PEI in DNA/PEI dispersion (N:P = 10) before and after the removal of the polyplexes, respectively

On the other hand, we evaluated the cytotoxicity of free PEI chains and the corresponding DNA/PEI dispersion on 293T cells using the MTT assay (Fig. 2.5). The initial cell viability of DNA/PEI at N:P = 3 is ~85 %, and is stably reduced as the increasing amount of free PEI chains, especially when $C_{PEI} \geq 2.7$ μg/mL, corresponding to N:P ≥ 10, clearing revealing that those free PEI chains are indeed the major cause of toxicity of DNA/PEI at high N:P ratios. It is also worth noting that in the general N:P range for transfection, i.e., N:P ≤ 10, both the DNA/PEI dispersion and free PEI chains exhibit relatively low cytotoxicity, with the cell viability well above 70 %.

Furthermore, we decided to add the 7 portions of free PEI chains at different times; namely, hours before or after the administration of the DNA/PEI polyplexes (N:P = 3), so that the final and total N:P ratio remains 10. We define t = 0 for the simultaneous addition of the DNA/PEI polyplexes (N:P = 3) and 7 portions of free PEI chains. Therefore, the negative time means that free PEI chains are added before the polyplex administration. Figure 2.6 shows that in the presence of free PEI chains, the transfection efficiency is typically $10^2$–$10^3$ times higher than that without free chains no matter whether these free polycations are introduced before or after the polyplex administration. To our knowledge, this is the first demonstration that the addition of free PEI chains even prior to the polyplexes can facilitate the gene transfection. It is clear that the simultaneous addition of the polyplexes (N:P = 3) and free polycationic chains (i.e., t = 0) leads to the highest

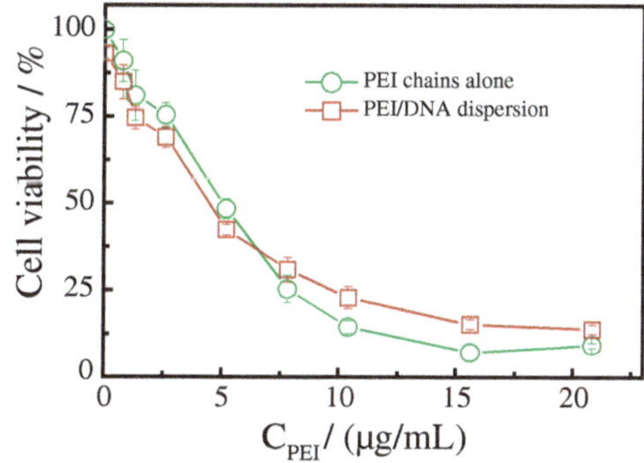


**Fig. 2.5** PEI concentration dependence of 293T cell viability in terms of free PEI chains alone and DNA/PEI dispersions, where $C_{PEI} = 2.7$ μg/mL is corresponding to N:P = 10

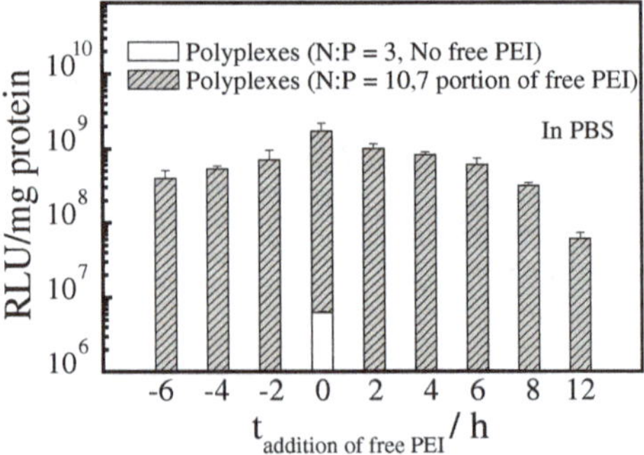


**Fig. 2.6** Effect of 7 portions of free PEI chains added at different times (prior to or post the administration of polyplexes formed at N:P = 3) on the gene transfection efficiency in 293T cells, where the total and final N:P ratio is kept to be 10 and the cell culture medium is not replaced before the addition of free PEI chains

transfection level, indicating a possibly cooperative role of the bound and free PEI chains in gene delivery.

Figure 2.6 also reveals that in the delayed addition of free PEI chains, i.e., $t > 0$, the gene transfection efficiency gradually decreases with the time interval between the administration of polyplexes and the addition of free PEI chains. To explain such a decrease, we have to find whether this decline is due to the insufficient incubation time since the incubation time was fixed to be 48 h, starting from the polyplex administration. The incubation-time dependence of the transgene expression, as shown in Fig. 2.7, clearly excludes such a possibility because the luciferase expression reaches its maximum at ~36 h after the polyplex administration. Therefore, the total 48-h incubation time is sufficient. The next question is whether these free PEI chains promote the cellular uptake in the extracellular space or other processes inside the cell. In the extracellular pathway, free PEI chains could facilitate the cellular internalization by increasing either the uptake rate or the endocytosis amount. To estimate the internalization timescale of the DNA/PEI dispersion, the cell culture medium with the polyplexes, if any, was replaced at different times after the polyplex administration. Figure 2.8 shows that the cellular uptake nearly ceases after ~6 h regardless of whether there are free PEI chains, but these free polycations enhance the final transfection levels by ~500 times even though they do not shorten the time required for internalization.

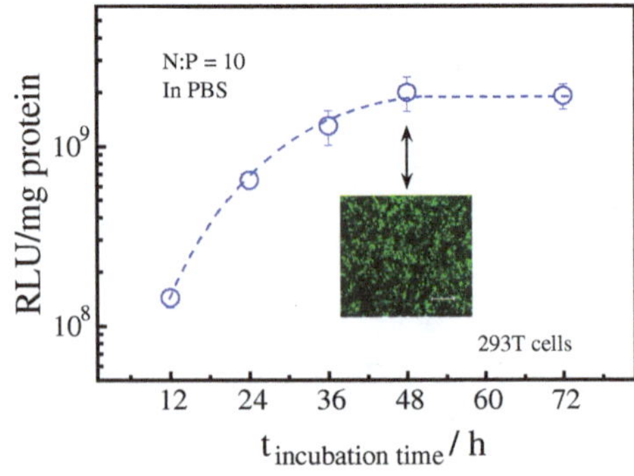


**Fig. 2.7** Incubation-time dependence of luciferase expression of DNA/PEI polyplexes formed at N:P = 10. *Inset* Fluorescent microscopic image of 293T cells transfected with pEGFP-N1 after 48 h. Scale bar: 200 μm

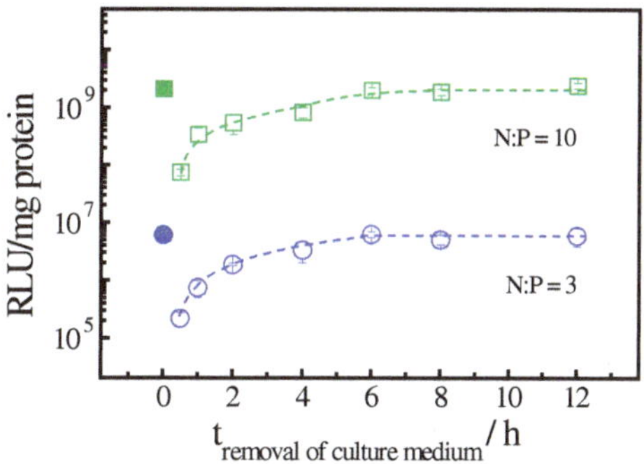


**Fig. 2.8** Effect of removal of cell culture medium on the gene transfection efficiency of DNA/PEI polyplexes formed in PBS with and without free PEI chains in the solution mixture, where "t = 0" means that no cell culture medium was replaced

Furthermore, flow cytometry was used to explore the internalization kinetics of DNA/PEI polyplexes and free PEI chains in the gene transfection. Generally, three characteristics can be extracted from the flow cytometry data; namely, (1) the fraction of cells containing Cy5 fluorescence (Fig. 2.9a); (2) the median of Cy5-DNA fluorescence intensity of Cy5-positive cell population ($F_{Cy5\text{-}DNA,\ Cy5+}$, Fig. 2.9b); and (3) the median of Cy5-DNA fluorescence intensity of total cell population ($F_{Cy5\text{-}DNA,\ tot}$, Fig. 2.9c). Figure 2.9a shows that in the presence of 7 portions of free PEI chains, more than 80 % of the cells have internalized polyplexes after a 30-min incubation, whereas the internalized polyplexes without free PEI chains are detected in only ~10 % of the cells. At 4 h post-incubation, ~95 % of the cells are Cy5-positive regardless of the addition of free PEI chains. In addition, $F_{Cy5\text{-}DNA,\ Cy5+}$, which is indicative of the polyplex amount per Cy5-positive cell, is stably 3–6 times higher during the first 6-h incubation when the extra 7 portions of PEI chains are applied (Fig. 2.9b). In combination, the addition of free PEI leads to a significant faster and remarkably increased internalization (Fig. 2.9c).

Moreover, different cellular uptake kinetics is revealed for polyplexes in the absence/presence of the free PEI chains (Fig. 2.10). When the polyplexes are applied alone, the number of cells they entered and their content per cell elevate simultaneously in the first 6 h of incubation. With the aid of 7 portions of free PEI chains, in contrast, the fraction of Cy5-positive cells quickly rises to ~80 % within just 0.5 h, and the enhanced internalization afterwards (1.5–6 h) is mainly attributed to the increasing amount of polyplexes inside one cell.

On the other hand, the 7 portions of free PEI chains, whether applied alone or with DNA/PEI polyplexes at N:P = 3, exhibit similar internalization kinetics (Fig. 2.11). Within just 30 min, these free PEI chains enter ~80 % of the cells, with their intracellular fluorescence intensity almost reaching the maximum plateau. Moreover, our CLSM result also confirms that the concentrations of free PEI chains at the cellular membrane and inside the cell quickly approach to the maximum ~30 min after they are added outside the cell (Fig. 2.12). This indicates that the free PEI chains have a significantly faster uptake rate, and perhaps a different internalization pathway, compared to the DNA/PEI polyplexes (~6 h). Notably, the major uptake timescale of free PEI (~0.5–1.5 h) is in good

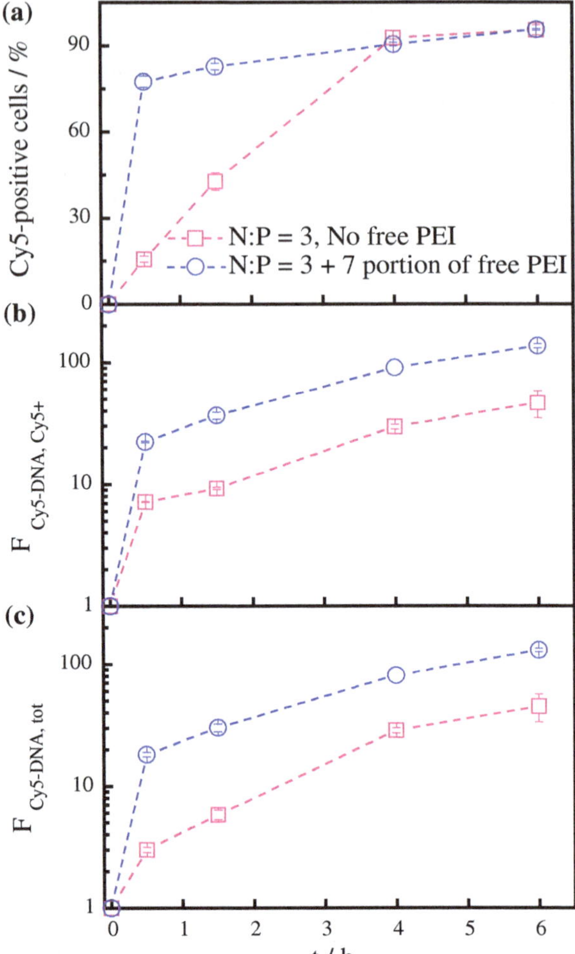


**Fig. 2.9** Effect of free PEI chains on the internalization of DNA/PEI polyplexes by flow cytometry. 293T cells were transfected with PEI/Cy5-DNA polyplexes alone (N:P = 3), or polyplexes plus 7 portions of free FITC-PEI (N:P = 10), harvested, and analyzed at the indicated time points. Internalization extent is expressed as **a** percentage of Cy5-positive cells; **b** median of Cy5-DNA fluorescence intensity of Cy5-positive cell population ($F_{Cy5\text{-}DNA,\ Cy5+}$); and **c** median of Cy5-DNA fluorescence intensity of total cell population ($F_{Cy5\text{-}DNA,\ tot}$)

accordance with the sharp increase in the number of polyplex-containing cells at N:P = 10 (Fig. 2.9a), suggesting that these excessive PEI chains might mainly play their roles in the first 1.5 h. It is still to be elucidated that how the free PEI chains promote the cellular internalization. They might either destabilize the cellular membrane to allow direct make the endocytosis of polyplexes more easily, or help the polyplexes to utilize an internalization pathway more favorable for intracellular trafficking [27, 28, 53].

To compare the contribution of free PEI chains in the extracellular and intracellular pathways, we added free PEI chains with or without removing the cell culture medium that contains the polyplexes (Fig. 2.13). Since the cellular internalization stops after ~6 h, those polyplexes in the extracellular space, if any,

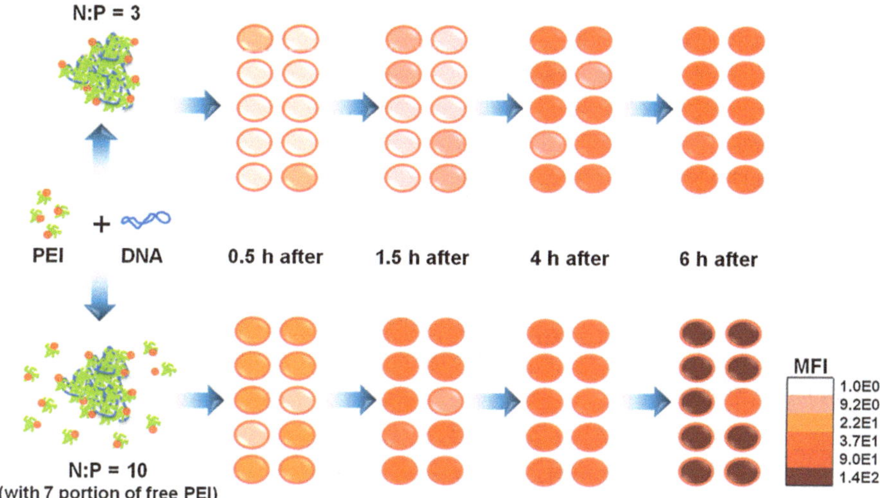


**Fig. 2.10** Schematic illustration of cellular internalization of DNA/PEI polyplexes (N:P = 3) in the absence and presence of 7 portions of free PEI chains, where different colors represents the median of DNA fluorescence intensity (MFI) inside one cell. Note that the depictions of different components do not reflect their actual sizes

should still remain there. For N:P = 10, when the polyplexes are removed, at 6 h post-incubation, before the addition of 7 portions of free PEI chains, the gene transfection efficiency is only attenuated by 2-fold in comparison with the non-removal group, indicating that those free PEI chains promote the cellular uptake a little. However, it is worth noting that this reduced transfection efficiency is still ~50-fold higher than that without the addition of free PEI chains (N:P = 3), clearly indicating that the free PEI chains mainly play their role inside the cells.

Finally, we explored the effect of FBS on the gene transfection efficiency of polyplexes alone (N:P = 3), or polyplexes with 7 portions of free PEI chains added at different times (N:P = 10). When the cell culture medium is supplemented with 10 % FBS, the initial transfection efficiency of polyplexes at N:P = 3 is reduced by ~5 times (Fig. 2.14a). Moreover, the 7 portions of free PEI chains added afterwards only increase the transfection level by 1–150 folds, whereas, in the serum-free medium, they can enhance the transfection efficiency by 10–300 times (Fig. 2.14b). This decreased efficacy is mainly attributed to the association of the cationic polyplexes and PEI chains with the negatively charged proteins in FBS; namely, it reduces the effective amount of free PEI chains, which indirectly supports our finding that free PEI chains plays a significant role in promoting the transfection. Therefore, for the future in vivo study, we have to consider how to incorporate free PEI chains together with polyplexes and keep their integrity in the blood stream so that they can be safely delivered to the targeted cells or organs.

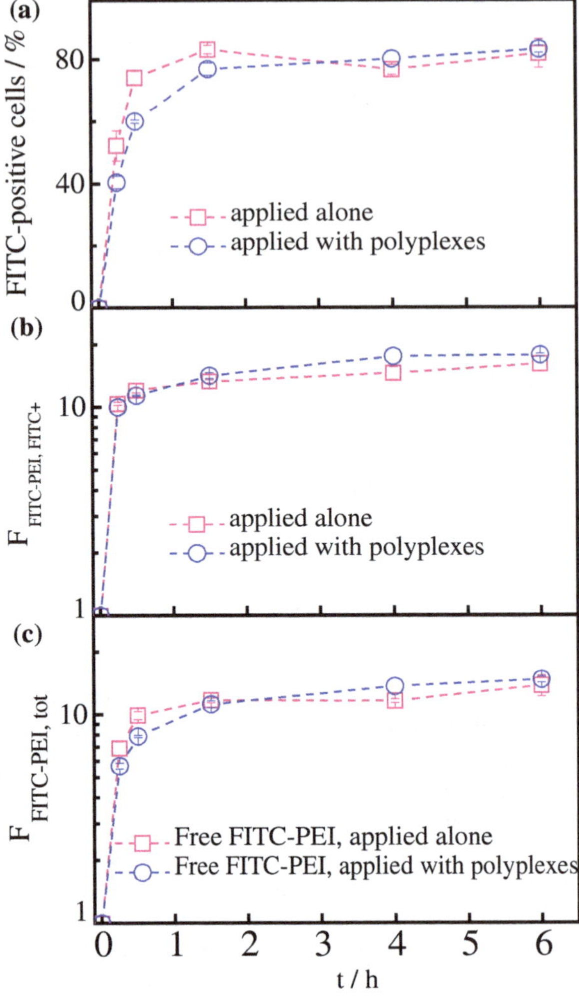


**Fig. 2.11** Comparison of cell internalization kinetics of free FITC-PEI chains applied alone or with DNA/PEI polyplexes of N:P = 3 in the gene transfection. Internalization extent is expressed as **a** percentage of FITC-positive cells; **b** median of FITC-PEI fluorescence intensity of FITC-positive cells ($F_{FITC\text{-}PEI,\ FITC+}$); and **c** median of FITC-PEI fluorescence intensity of total cell population ($F_{FITC\text{-}PEI,\ tot}$)

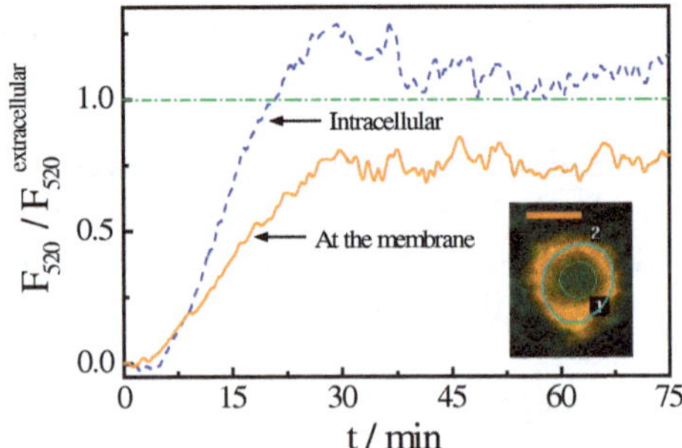


**Fig. 2.12** Cellular membrane translocation kinetics of FITC-labeled *b*PEI-25 K chains, where the mean FITC-PEI fluorescence intensities at the membrane (on the circle 2) and inside the cell (inside circle 1, scale bar: 10 μm) are normalized by that outside the cell

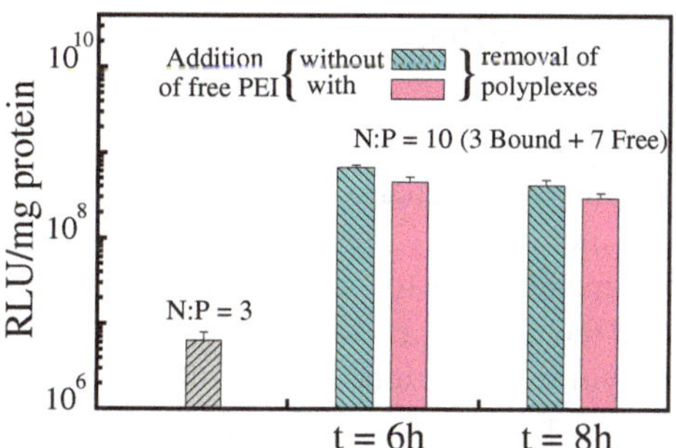


**Fig. 2.13** Effect of free PEI chains on the intracellular trafficking of DNA/PEI polyplexes formed in PBS, where 7 portions of free PEI chains were added, respectively, at 6 h and 8 h post-administration of the polyplexes (N:P = 3), with or without the removal of cell culture medium

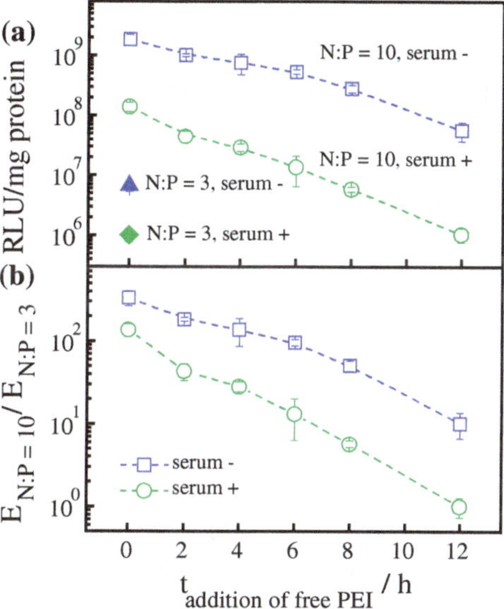


**Fig. 2.14** **a** Effect of FBS (10 % in the cell culture medium) on the transfection efficiency of polyplexes alone (N:P = 3), or polyplexes with 7 potions of free PEI added at different times (N:P = 10). **b** Relative transfection efficiency of polyplexes plus free PEI ($E_{N:P\ =\ 10}$) to polyplex alone ($E_{N:P\ =\ 3}$) in the serum-free (−) and serum-containing (+) medium, respectively. The cell culture medium is not replaced before the addition of free PEI chains

In sum, our results reveal the existence of some cooperative actions between the polyplexes and free polycationic chains. The cationic PEI chains inside each polyplex condense anionic DNA chains into a small and compact particle so that the endocytosis is facilitated. These bound PEI chains also shield DNA from the intracellular degradation; while those free PEI chains in the solution mixture mainly navigate some unknown intracellular trafficking barriers, including endolysosomal formation and/or escape and the subsequent nuclear localization. It is a remaining challenge to map these intracellular trafficking pathways and understand how free PEI chains exactly promote the gene transfection inside the cell by using various existing tools of molecular biology.

## 2.4 Conclusion

A combination of LLS and gel electrophoresis results confirms that most of anionic DNA chains are complexed and condensed by cationic PEI chains when the molar ratio of nitrogen from PEI to phosphate from DNA (N:P) reaches $\sim 3$, independent of the PEI chain length and solvent (pure water or PBS), revealing that the charge neutrality (electrostatic interaction) is the main driving force for the polyplex formation. In the solution mixture with N:P $> 3$, there are two kinds of PEI chains: bound to DNA and free in the solution mixture. The bound PEI chains inside the polyplexes provide charge compensation so that DNA is condensed and protected from degradation. Our current study convincingly reveals that it is those free PEI chains, rather than other physical properties of the DNA/PEI polyplexes, that play a vital role in promoting the gene transfection. The addition of free PEI leads to a significant faster and more efficient cellular uptake of polyplexes, but these free PEI chains mainly contribute to the subsequent intracellular trafficking. Our finding leads to a different thinking in the development of non-viral vectors; namely, we might not need to invest much of our effort on the synthesis of different cationic polymers, but focus on how the free polycationic chains promote the intracellular trafficking of the polyplexes. As for in vivo experiments, we have to consider how to incorporate free chains together with polyplexes and keep their integrity during circulation so that they can be safely delivered to the targeted cells or organs.

## References


1. Check, E. (2002). *Nature, 420*, 116–118.
2. Raper, S. E., Chirmule, N., Lee, F. S., Wivel, N. A., Bagg, A., Gao, G. P., et al. (2003). *Molecular Genetics and Metabolism, 80*, 148–158.
3. Hacien-Bey-Abina, S. (2003). *Science, 302*, 568–568.
4. Pack, D. W., Hoffman, A. S., Pun, S., & Stayton, P. S. (2005). *Nature Reviews Drug Discovery, 4*, 581–593.
5. Li, S. D., & Huang, L. J. (2007). *Journal of Controlled Release, 123*, 181–183.
6. Mintzer, M. A., & Simanek, E. E. (2009). *Chemical Reviews, 109*, 259–302.
7. Boussif, O., Lezoualch, F., Zanta, M. A., Mergny, M. D., Scherman, D., Demeneix, B., et al. (1995). *Proceedings of the National Academy of Sciences of the United States of America, 92*, 7297–7301.
8. Lungwitz, U., Breunig, M., Blunk, T., & Gopferich, A. (2005). *European Journal of Pharmaceutics and Biopharmaceutics, 60*, 247–266.
9. Neu, M., Fischer, D., & Kissel, T. J. (2005). *Gene Medicine, 7*, 992–1009.
10. Breunig, M., Lungwitz, U., Liebl, R., & Goepferich, A. (2007). *Proceedings of the National Academy of Sciences of the United States of America, 104*, 14454–14459.
11. Deng, R., Yue, Y., Jin, F., Chen, Y. C., Kung, H. F., Lin, M. C. M., et al. (2009). *Controlled Release, 140*, 40–46.
12. Ogris, M., Brunner, S., Schuller, S., Kircheis, R., & Wagner, E. (1999). *Gene Therapy, 6*, 595–605.

13. Cheng, H., Zhu, J. L., Zeng, X., Jing, Y., Zhang, X. Z., & Zhuo, R. X. (2009). *Bioconjugate Chemistry, 20*, 481–487.
14. Zanta, M. A., Boussif, O., Adib, A., & Behr, J. P. (1997). *Bioconjugate Chemistry, 8*, 839–844.
15. Diebold, S. S., Kursa, P., Wagner, E., Cotten, M., & Zenke, M. J. (1999). *Biological Chemistry, 274*, 19087–19094.
16. Pollard, H., Remy, J. S., Loussouarn, G., Demolombe, S., Behr, J. P., & Escande, D. J. (1998). *Biological Chemistry, 273*, 7507–7511.
17. Godbey, W. T., Wu, K. K., & Mikos, A. G. (1999). *Proceedings of the National Academy of Sciences of the United States of America, 96*, 5177–5181.
18. Bieber, T., Meissner, W., Kostin, S., Niemann, A., & Elsasser, H. P. J. (2002). *Controlled Release, 82*, 441–454.
19. Sonawane, N. D., Szoka, F. C., & Verkman, A. S. J. (2003). *Biological Chemistry, 278*, 44826–44831.
20. Suh, J., Wirtz, D., & Hanes, J. (2003). *Proceedings of the National Academy of Sciences of the United States of America, 100*, 3878–388.
21. Akinc, A., Thomas, M., Klibanov, A. M., & Langer, R. J. (2005). *Gene Medicine, 7*, 657–663.
22. Kulkarni, R. P., Wu, D. D., Davis, M. E., & Fraser, S. E. (2005). *Proceedings of the National Academy of Sciences of the United States of America, 102*, 7523–7528.
23. de Bruin, K., Ruthardt, N., von Gersdorff, K., Bausinger, R., Wagner, E., Ogris, M., et al. (2007). *Molecular Therapy, 15*, 1297–1305.
24. de Bruin, K. G., Fella, C., Ogris, M., Wagner, E., Ruthardt, N., & Brauchle, C. J. (2008). *Controlled Release, 130*, 175–182.
25. Gabrielson, N. P., & Pack, D. W. J. (2009). *Controlled Release, 136*, 54–61.
26. Won, Y–. Y., Sharma, R., & Konieczny, S. F. J. (2009). *Controlled Release, 139*, 88–93.
27. Lee, H., Kim, I. K., & Park, T. G. (2010). *Bioconjugate Chemistry, 21*, 289–295.
28. Wagner, E., Cotten, M., Foisner, R., & Birnstiel, M. L. (1991). *Proceedings of the National Academy of Sciences of the United States of America, 88*, 4255–4259.
29. Godbey, W. T., Wu, K. K., & Mikos, A. G. (1999). *Journal of Biomedical Materials Research, 45*, 268–275.
30. Wightman, L., Kircheis, R., Rossler, V., Carotta, S., Ruzicka, R., Kursa, M., et al. (2001). *Gene Medicine, 3*, 362–372.
31. Kunath, K., von Harpe, A., Fischer, D., Peterson, H., Bickel, U., Voigt, K., et al. (2003). *Controlled Release, 89*, 113–125.
32. Honore, I., Grosse, S., Frison, N., Favatier, F., Monsigny, M., & Fajac, I. J. (2005). *Controlled Release, 107*, 537–546.
33. Pun, S. H., Bellocq, N. C., Liu, A. J., Jensen, G., Machemer, T., Quijano, E., et al. (2004). *Bioconjugate Chemistry, 15*, 831–840.
34. Ooya, T., Choi, H. S., Yamashita, A., Yui, N., Sugaya, Y., Kano, A., et al. (2006). *Journal of the American Chemical Society, 128*, 3852–3853.
35. Tian, H. Y., Xiong, W., Wei, J. Z., Wang, Y., Chen, X. S., Jing, X. B., et al. (2007). *Biomaterials, 28*, 2899–2907.
36. Sun, Y. X., Xiao, W., Cheng, S. X., Zhang, X. Z., & Zhuo, R. X. J. (2008). *Controlled Release, 128*, 171–178.
37. Liang, B., He, M. L., Chan, C. Y., Chen, Y. C., Li, X. P., Li, Y., et al. (2009). *Biomaterials, 30*, 4014–4020.
38. Bloomfield, V. A. (1997). *Biopolymers, 44*, 269–282.
39. Lai, E., & van Zanten, J. H. (2001). *Biophysical Journal, 80*, 864–873.
40. Vuorimaa, E., Urtti, A., Seppanen, R., Lemmetyinen, H., & Yliperttula, M. (2008). *Journal of the American Chemical Society, 130*, 11695–11700.
41. Ziebarth, J., & Wang, Y. M. (2009). *Biophysical Journal, 97*, 1971–1983.
42. Behr, J. P. (1997). *Chimia, 51*, 34–36.
43. Clamme, J. P., Azoulay, J., & Mely, Y. (2003). *Biophysical Journal, 84*, 1960–1968.

44. Boeckle, S., von Gersdorff, K., van der Piepen, S., Culmsee, C., Wagner, E., & Ogris, M. J. (2004). *Gene Medicine, 6*, 1102–1111.
45. Fahrmeir, J., Gunther, M., Tietze, N., Wagner, E., & Ogris, M. J. (2007). *Controlled Release, 122*, 236–245.
46. Clamme, J. P., Krishnamoorthy, G., & Mely, Y. (2003). *Biochimica et Biophysica Acta, Biomembranes, 1617*, 52–61.
47. Saul, J. M., Wang, C. H. K., Ng, C. P., & Pun, S. H. (2008). *Advanced Materials, 20*, 19–25.
48. Berne, B., & Pecora, R. (1976). *Dynamic light scattering*. New York: Plenum Press.
49. Chu, B. (1991). *Laser light scattering* (2nd ed.). New York: Academic Press.
50. Hunter, R. J. (2000). *Foundations of colloid science* (2nd ed.). Oxford: Oxford Press.
51. Forrest, M. L., & Pack, D. W. (2002). *Molecular Therapy, 6*, 57–66.
52. Breunig, M., Lungwitz, U., Liebl, R., Fontanari, C., Klar, J., Kurtz, A., et al. (2005). *Journal of Gene Medicine, 7*, 1287–1298.
53. Rejman, J., Bragonzi, A., & Conese, M. (2005). *Molecular Therapy, 12*, 468–474.

# Chapter 3
# Revisiting Complexation Between DNA and Polyethylenimine: The Effect of Length of Free Polycationic Chains on Gene Transfection

## 3.1 Introduction

The gene therapy, considered as the treatment of genetically-caused diseases by transferring exogenous nucleic acids into specific cells of patients, has attracted great interests over the past few decades [1]. It has been gradually realized that the development of safe, efficient and controllable gene-delivery vectors has become a bottleneck in clinical applications. The gene transfection vectors can be generally divided as viral and non-viral ones. The non-viral vectors, including cationic polymers and lipids, offer several advantages in comparison with the viral ones, such as low oncogenicity, construction flexibility and facile fabrication [1–3]. However, their up-to-now low efficacy greatly limits their potential clinical applications. The introduction of polyethylenimine (PEI) as a non-viral vector represented a big leap because of its much higher efficiency in facilitating the gene transfection [4–6]. After such a discovery, PEI has been modified in a myriad of ways to reduce its high cytotoxicity by shielding its cationic charges, enhances its efficiency by coupling short PEI chains with intracellular biodegradable linkers [7, 8], increase its serum stability by grafting poly (ethylene glycol) (PEG) on it [9, 10], and improve its cell targeting by attaching some functional molecules on its surface [9–12]. However, less attention has been paid to even ask why PEI remains one of the best non-viral vectors, how it promotes the intracellular trafficking of the polyplexes, and when and where DNA is released from the polyplexes [13–24].

The first step in the preparation of non-viral vectors is to package long anionic DNA chains into a small particle, i.e., the DNA complexation and encapsulation. Previous studies have accumulated many data about the effects of size, density and zeta-potential of the polyplexes on their final transfection efficiency [25–28], but a coherent picture is still lacking. Previous studies have attributed the high trans fection efficiency of PEI to the so-called "proton sponge" effect; namely, further protonation of the PEI chains inside the endolysosomes due to their primary and secondary amines, which should lead to an influx of counter (chloride) ions so that the raising of the osmotic pressure inside could burst the endocytic vesicle and release the polyplexes [4, 29]. Many people, especially those who jumped into this

Y. Yanan, *How Free Cationic Polymer Chains Promote Gene Transfection*, Springer Theses, DOI: 10.1007/978-3-319-00336-8_3,

research field later, have taken this mechanism as granted. However, a number of researchers have always questioned whether such a "proton sponge" effect plays a dominant role in the high efficiency of PEI because of some contradictive results [23, 30]. For example, long PEI chains are much more effective than short ones; and on the other hand, if only considering the colligativity, we know from thermodynamics that for a given weight concentration (g/mL), both short and long PEI chains should lead to a very similar number of chloride influx but short chains themselves should generate a slightly higher osmotic pressure inside the endocytic vesicles. Therefore, one could not put these two well-known facts together.

Recently, using a combination of different methods to characterize the size, molar mass and surface charge of the DNA/PEI polyplexes formed under different conditions, we previously found that nearly all the DNA chains are condensed by PEI to form the DNA/PEI polyplexes when the molar ratio of nitrogen from PEI to phosphate from DNA (N:P) reaches $\sim$3, but the polyplexes have a high in vitro gene transfection efficiency only when N:P $\geq$ 10 [31]. Putting these two facts together, we concluded and confirmed that (1) the extra 7 portions of the PEI chains are free in the solution mixture; and (2) it is these 7-portion free PEI chains that greatly promote the gene transfection no matter whether they are applied hours before or after the administration of the polyplexes (N:P = 3). Therefore, there are two kinds of PEI chains in the solution mixture: bound to DNA and free in the solution. Note that such an effect of free PEI chains was reported before but most of the results were only qualitative [32–36].

In the current study, we have further evaluated the effect of the length of both the bound and free PEI chains on the gene transfection by using different quantitative experimental methods. Particularly, we have explored how the chain length affects the cellular uptake and the subsequent intracellular trafficking. More importantly, we have investigated how the free chains with different lengths possibly prevent the development of the later endolysosomes and help the release of the polyplexes from the endosomes. In addition, we have also studied the effects of the chain topology and chemical structure.

## 3.2 Materials and Methods

### *3.2.1 Materials and Cell Lines*

Three branched PEIs ($M_w$ = 800, 2,000, and 25,000 g/mol, denoted as *b*PEI-0.8K, *b*PEI-2K, and *b*PEI-25K) and two linear PEIs ($M_w$ = 2,500 and 25,000 g/mol, denoted as *lin*PEI-2.5K and *lin*PEI-25K) were respectively purchased from Sigma-Aldrich and Polysciences, and used without further purification. FITC-labeled *b*PEI-25K was prepared as described [31]. Initial plasmid DNA pGL3-control vector (5,256 bp) encoding modified firefly luciferase was purchased from Promega (USA). A large amount of this plasmid was prepared by ourselves using a Qiagen Plasmid Maxi Kit (Qiagen, Germany). 3-(4, 5-dimethylthiazol-2-yl)-2, 5-diphenyltetrazolium bromide (MTT) and bafilomycin A1 (Baf-A1) were purchased from Sigma-Aldrich (Deutschland). POPO-3 iodide was purchased from Invitrogen (USA). The Bright-Glo assay and CytoTox 96 non-radioactive cytotoxicity assay kits were purchased from Promega (USA). The fetal bovine serum (FBS), Dulbecco's modified Eagle's medium (DMEM) and penicillin–streptomycin were products of GIBCO (USA). 293T cells were grown at 37 °C, 5 % $CO_2$ in DMEM supplemented with 10 % FBS, penicillin at 100 units/mL and streptomycin at 100 μg/mL.

### *3.2.2 Formation of DNA/PEI Polyplexes*

The plasmid DNA (*p*DNA) was complexed with different PEIs in either distilled water or phosphate buffered saline (PBS) to form the DNA/PEI polyplexes as follows. Different amounts of the PEI solution ($C = 10^2$–$10^3$ μg/mL) were added dropwise into a dilute DNA solution ($C = 14.5$ μg/mL), resulting in different molar ratios of nitrogen from PEI to phosphate from DNA (N:P). Each resultant dispersion was incubated for 5 min at room temperature before being added to the cell culture medium. The DNA/PEI polyplexes were analyzed by the gel-shift assay (110 V for 20 min), in which the polyplexes were mixed with 6 × loading buffer and then loaded on a 0.8 % (w/v) agarose gel containing ethidium bromide in TBE buffer and the result was photographed under UV. The DNA binding was also characterized using laser light scattering.

### *3.2.3 Formation of Phospholipid Vesicles*

Chloroform solutions of soybean phospholipids (0.09 g, injectable grade, Shanghai Taiwei Incorporation) and cholesterol (0.03 g, Chengdu Kelong Incorporation) were mixed in a 25 mL round-bottom flask and the chloroform was removed at room temperature by rotary evaporator to form a uniform film. The flask was placed under vacuum for an additional 6 h. The dried film was hydrated by vortex with 20 mL of warm DI water and then subjected forty ultrasonic rounds (a total of 10 min) using an ultrasonic cell crusher machine. The solution mixture was further subjected to extrusion through a 450-nm PTFE hydrophilic filter to remove the large vesicles. Vesicle solutions were stored at 4 °C and diluted as needed for laser light scattering characterization and zeta-potential measurement.

### *3.2.4 POPO-3 Exclusion Assay*

*p*DNA was labeled with POPO-3 dye at a ratio of 1 dye per 100 base pairs. The DNA/PEI dispersion with different chosen N:P ratios were prepared as described above. The final DNA concentration was kept a constant (1.6 μg DNA per 100 μL PBS). After ~5-min incubation at room temperature, the relative fluorescence intensity ($F$) of each solution mixture was recorded using a Hitachi F7000 fluorescence spectrophotometer (excitation 534 nm, emission 568 nm). The fraction of the uncomplexed DNA chains ($x$) free in the solution mixture was determined by $x = (F_{\text{polyplex}} - F_{\text{blank}})/(F_{\text{DNA}} - F_{\text{blank}})$.

### 3.2.5 Laser Light Scattering

A commercial LLS instrument (ALV5000) with a vertically polarized 22-mV He–Ne laser (632.8 nm, Uniphase) was used to characterize the polyplexes mediated by different PEIs. In static LLS [37], we can obtain the weight-averaged molar mass ($M_w$) and the z-averaged root-mean square radius of gyration ($\langle R_g \rangle_z$) of scattering objects in a sufficiently dilute solution/dispersion from the angular and concentration dependence of the excess absolute scattering intensity (Rayleigh ratio $R_{vv}(q)$) as

$$\frac{KC}{R_{vv}(q)} \cong \frac{1}{M_w}\left(1 + \frac{1}{3}q^2R_g^2\right) \tag{3.1}$$

where $K \equiv 4\pi^2(dn/dC)^2/(N_A\lambda_0^4)$ and $q \equiv (4\pi n/\lambda_0)\sin(\theta/2)$ with $dn/dC$, $N_A$, $\lambda_0$, $n$ and $\theta$ the specific refractive index increment, the Avogadro number, the incident wavelength in vacuum, the refractive index of solvent, and the scattering angle, respectively. Note that we have neglected the concentration correction here. The values of $(dn/dC)_{632.8\text{ nm}}$ of DNA and different PEIs in water were determined using a novel laser differential refractometer [38]. $(dn/dC)_{632.8\text{ nm}}$ of polyplexes was calculated by using $(dn/dC)_{\text{polyplex}} = w_{\text{DNA}}(dn/dC)_{\text{DNA}} + w_{\text{PEI}}(dn/dC)_{\text{PEI}}$, where $w_{\text{DNA}}$ and $w_{\text{PEI}}$ are two normalized weight fractions of DNA and PEI inside the polyplexes, i.e., $w_{\text{DNA}} + w_{\text{PEI}} = 1$. The measurable angular range was 20–50° for the polyplexes to ensure that $qR_g < 1$. All of the measurements were carried out at 25.0 ± 0.1 °C.

In dynamic LLS [39], the Laplace inversion of each measured intensity–intensity time correlation function ($G^{(2)}(q, t)$) in the self-beating mode can be related to a line-width distribution $G(\Gamma)$. For a diffusive relaxation, $\Gamma$ is further related to the translational diffusion coefficient $D$ by $(\Gamma/q^2)_{C \to 0,\, q \to 0} = D$. Therefore, $G(\Gamma)$ can be converted into a translational diffusion coefficient distribution $G(D)$ or a hydrodynamic radius distribution $f(R_h)$ using the Stokes-Einstein equation, $R_h = k_BT/(6\pi\eta D)$, where $k_B$, $T$, and $\eta$ are the Boltzmann constant, the absolute temperature, and the solvent viscosity, respectively.

### 3.2.6 Zeta-Potential Measurement

The average mobility ($\mu_E$) of the polyplexes under an electric field in an aqueous solution was determined from the frequency shift in a laser Doppler spectrum using a commercial zeta-potential spectrometer (ZetaPlus, Brookhaven) with two platinum-coated electrodes and one He–Ne laser as the light source. Each data point presented in the mobility measurement was averaged over 10 times at 25 °C. The zeta-potential ($\zeta_{\text{potential}}$) can be calculated from $\mu_E$ using $\mu_E = 2\varepsilon\zeta_{\text{potential}} f(\kappa R_b)/(3\eta)$, where $\varepsilon$ is the permittivity of water and $1/\kappa$ is the Debye screening length [40]. When $\kappa R_b << 1$ (the Hückel limit), $f(\kappa R_b) \approx 1$; while $\kappa R_b >> 1$ (the Smoluchowski limit), $f(\kappa R_b) \approx 1.5$. In the current study, the Hückel and Smoluchowski limits are respectively used to calculate $\zeta_{\text{potential}}$ in water and PBS.

### 3.2.7 In Vitro Gene Transfection

The in vitro gene transfection efficiency was quantified by using the luciferase transfection assays, in which plasmid pGL3 was used as an exogenous reporter gene. 293T cells were seeded in a 48-well plate at an initial density of 120,000 cells per well, 24 h prior to the gene transfection. Each DNA/PEI dispersion with a desired N:P ratio was further diluted in serum-free medium and then administered to the cells at a final concentration of 0.4 μg DNA per well. The complete DMEM medium (600 μL/well) was added 6 h after the administration of the DNA/PEI polyplexes. Using a GloMax 96 microplate luminometer (Promega, USA) and the Bio-Rad protein assay reagent, we respectively determined the transgene expression level and the corresponding protein concentration in each well at 48 h post-administration of the polyplexes. The gene transfection efficiency is expressed as a relative luminescence unit (RLU) per cellular protein (mean ± SD of triplicates).

### 3.2.8 Cytotoxicity Assay

The cytotoxicity of free PEI chains and the corresponding DNA/PEI dispersions was evaluated on 293T cells by using the MTT assay. 293T cells were seeded in a 96-well plate at an initial density of 60,000 cells per well. After 24 h, free PEI chains alone or with DNA to form the PEI/DNA dispersions were respectively added to cells at different chosen concentrations and N:P ratios. For DNA/PEI dispersion, the final DNA concentration is 0.2 μg/well in a total volume of 100 μL. The treated cells were incubated in a humidified environment with 5 % $CO_2$ at 37 °C for 48 h. The MTT reagent (in 20 μL PBS, 5 mg/mL) was then added to each well. The cells were further incubated for 4 h at 37 °C. The medium in each well was then removed and replaced by 100 μL of DMSO. The plate was gently agitated for 15 min before the absorbance ($A$) at 490 nm was recorded by a microplate reader (Bio-rad, USA). The cell viability ($y$) was calculated by $y = (A_{treated}/A_{control}) \times 100\ \%$, where $A_{treated}$ and $A_{control}$ are the absorbance of the cells cultured with polymer/polyplex and fresh culture medium, respectively. Each experiment condition was done in quadruplicate. The data was shown as the mean value plus a standard deviation (±SD).

### *3.2.9 Cellular Uptake of Polyplexes by Flow Cytometry*

Plasmid pGL3 was covalently labeled with the fluorophore Cy5 using a Label IT nucleic acid labeling kit (Mirus, Madison, WI) as the manufacturer's instructions. The gel electrophoresis results confirmed that the labeling has no visible interference on the formation of DNA/PEI polyplexes (data not shown). 293T cells were seeded in a 12-well plate at an initial density of 500,000 cells per well. After 24 h, Cy5-pGL3/*b*PEI-25K polyplexes (N:P = 3) without and with 7 portions of free *b*PEI-2K or *b*PEI-25K chains (N:P = 10) were added to the 293T cells in serum-free DMEM. The cells were incubated at 37 °C and harvested after different desired incubation times. The harvested cells were briefly rinsed twice with PBS containing 0.001 % SDS and then with only PBS to remove the polyplexes remained outside of the cells [41]. The cells were further detached by 0.05 % trypsin/EDTA supplemented with 20 mM sodium azide to prevent further endocytosis [42]. Finally, the cells were washed twice by pelleting and then resuspending in ice-cold PBS containing 2 % FBS. The confocal microscopic images of the cells were taken to ensure that the removal of the polyplexes from the extracellular space and the cell surface was complete (data not shown). The cellular uptake of polyplexes was assayed by using a FC 500 flow cytometry system (Beckman Coulter, USA). The fluorophore Cy5 was excited at 635 nm and detected at 675/15 nm. To discriminate the viable cells from dead ones and to exclude possible doublets, the cells were gated by the forward/side scattering and pulse width. 10,000 gated events per sample were collected at least in duplicate.

### *3.2.10 Measurement of Intracellular pH Around Polyplexes*

Plasmid pGL3 was covalently double-labeled with pH-sensitive FITC and pH-insensitive Cy5 using a Label IT nucleic acid labeling kit (Mirus, Madison, WI). 293T cells were seeded in a 6-well plate at an initial density of 1,000,000 cells per well. After 24 h, double-labeled pGL3/*b*PEI-25K polyplexes (N:P = 3) without and with 7 portions of free *b*PEI-2K or *b*PEI-25K chains (N:P = 10) were added to the 293T cells in serum-free DMEM, respectively. After 6-h incubation at 37 °C, the cells were harvested as described before and pelleted into six separate eppendorf vials. The cells in two vials were resuspended with PBS containing 2 %

FBS, whereas the cells in the rest four vials were respectively resuspended with four intracellular pH clamping buffers (pH = 5.33, 6.03, 6.73 and 7.43) [43]. The cells in each vial were further washed by pelleting and resuspending in an appropriate and corresponding buffer. The FITC/Cy5 fluorescence ratio of each sample was measured using the flow cytometer. The fluorophores FITC and Cy5 were excited at 488 nm and 635 nm, respectively. The corresponding emissions were detected at 525/10 and 675/15 nm, respectively. The FITC/Cy5 fluorescence ratio was calculated using the median values of the FITC and Cy5 fluorescence intensities of the total cell population (10,000 cells). The four intracellular pH clamped samples generate a linear pH calibration, from which the FITC/Cy5 ratio of each of the other two duplicate samples corresponds to an average pH around the DNA/PEI polyplexes.

### *3.2.11 Lactate Dehydrogenase Membrane Integrity Assay*

293T cells were seeded in a 96-well plate at an initial density of 60,000 cells per well. After 24 h, cells were treated with *b*PEI-2K or *b*PEI-25K chains at different desired concentrations. The treated cells were incubated in a humidified environment with 5 % $CO_2$ at 37 °C for 6 h. To determine the maximum Lactate Dehydrogenase (LDH) release, the 10 × lysis solution was added to the control group 2 h prior to the usage of a CytoTox 96 Non-radioactive cytotoxicity assay kit (Promega, USA). In each measurement, 50 μL of the cell culture solution from each well of the plate was aspirated and mixed with 50 μL of the reconstituted substrate mix in a new 96-well plate. After 30-min incubation at the room temperature, 50 μL of the stop solution was added to each well before the absorbance ($A$) at 490 nm was recorded by a microplate reader (Bio-Rad, USA). The degree of LDH Release, defined as $(A_{\text{treated}} - A_{\text{control}})/(A_{\text{maximum}} - A_{\text{control}}) \times 100\,\%$, represents the membrane disruption induced by different PEIs, where $A_{\text{treated}}$ and $A_{\text{control}}$ are the absorbance values of the cells cultured with and without PEI, and $A_{\text{maximum}}$ is the absorbance value of the cells in the maximum LDH release group. Each experiment condition was done in quadruplicate. The data were shown as the mean value plus a standard deviation (±SD).

## 3.3 Results and Discussion

Figure 3.1 shows the complexation profiles of DNA and branched PEI with different chain lengths, respectively evaluated by the gel-shift and POPO-3 exclusion assays. Progressive condensation can be revealed from the retarded mobility of the DNA bands and their reduced fluorescence intensity (Fig. 3.1a). Clearly, the two DNA bands (supercoiled and relaxed forms) disappear when N:P ~ 2–3, independent of the PEI chain length. Further, we quantitatively estimated the extent of

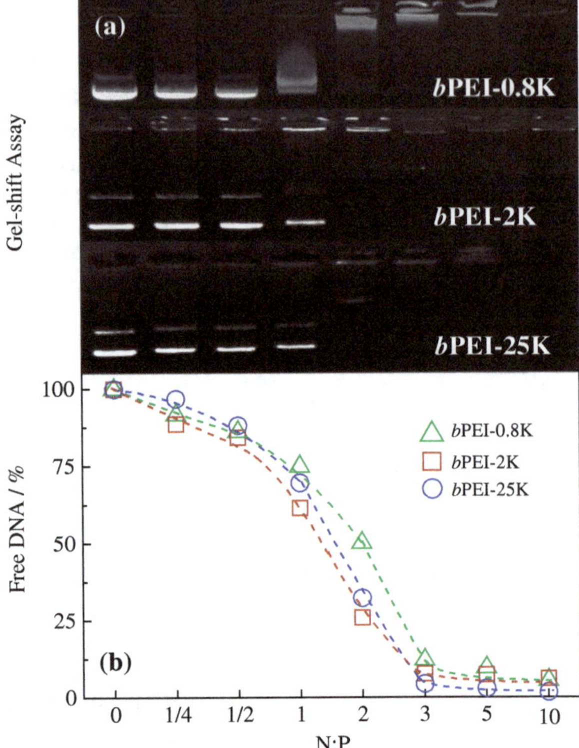


**Fig. 3.1** Complexation of DNA by PEI with different chain lengths, evaluated by **a** gel-shift assay and **b** POPO-3 exclusion assay

free and uncomplexed DNA chains at each given N:P ratio by using a double-strand DNA intercalating dye, POPO-3, whose fluorescence emission is dramatically enhanced upon its binding to DNA. The complexation of DNA with PEI excludes POPO-3 so that the fluorescence intensity decreases. Figure 3.1b shows that over 90 % of the DNA chains have been condensed into the polyplexes at N:P $\sim$ 3, irrespective of the PEI chain length.

On the other hand, we explored the N:P ratio dependence of the weight-averaged molar mass ($M_w$), the average hydrodynamic radius ($\langle R_h \rangle$), the average chain density($\langle \rho \rangle \equiv M_w/[N_A(4/3)\pi \langle R_h \rangle]$), and the zeta-potential ($\zeta_{potential}$) of the DNA/PEI complexes with different PEI chains using a combination of laser light scattering (LLS) and zeta-potential measurement (Fig. 3.2). There is no significant difference among the complexation profiles of DNA and PEI with different chain lengths. Namely, the polyplexes have a reversed positive $\zeta_{potential}$ at N:P $\sim$ 2–2.5 and further addition of more PEI chains in the solution mixture (i.e., N:P > 3) has nearly no effect on $M_w$, $\langle R_h \rangle$, $\langle \rho \rangle$ and $\zeta_{potential}$, revealing that those PEI chains added after N:P > 3 are free in the solution mixture, which are invisible in LLS because the scattered light intensity is proportional to the square of mass of a scattering object. It is worth noting that short chains (*b*PEI-0.8K) lead to larger DNA/PEI polyplexes ($M_w \sim 2.5 \times 10^8$ g/mol and $\langle R_h \rangle \sim 110$ nm) but

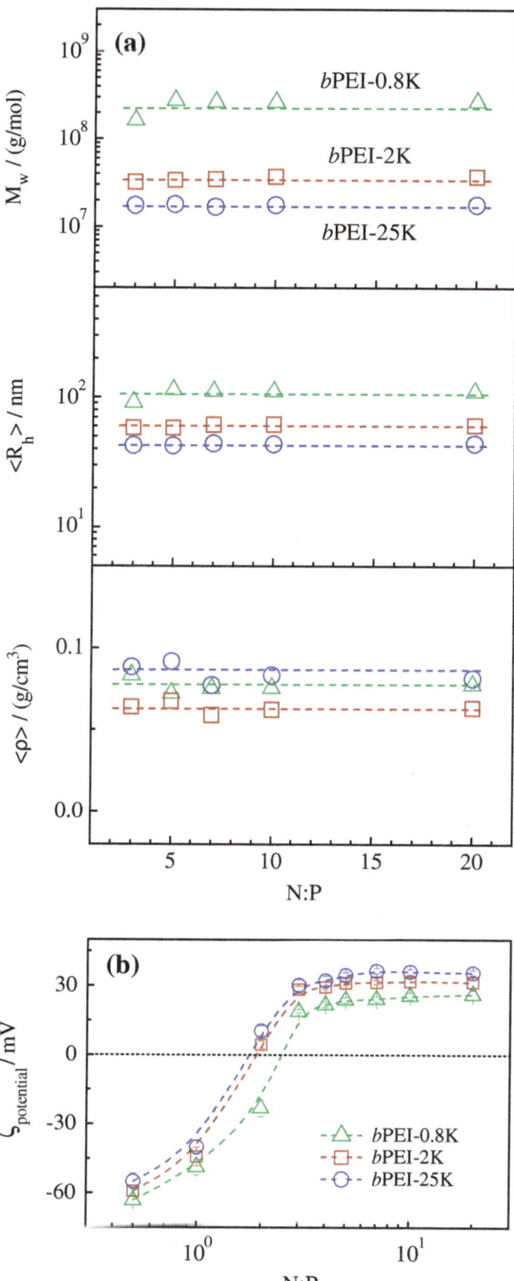


**Fig. 3.2** **a** N:P ratio dependence of weight-averaged molar mass ($M_w$), average hydrodynamic radius ($<R_h>$) and average chain density ($<\rho> \equiv M_w/[N_A(4/3)\pi <R_h>^3]$) of different PEI-mediated polyplexes formed in water. **b** N:P ratio dependence of zeta-potential ($\zeta_{potential}$) of different PEI-mediated polyplexes in water

a lower $\zeta_{potential}$ ($\sim$ 25 mV) than long chains (*b*PEI-25K, $M_w \sim 1.8 \times 10^7$ g/mol, $<R_h> \sim 45$ nm, and $\zeta_{potential} \sim 35$ mV), presumably due to the previously proposed viscoelastic effect [44, 45]. For comparison, we also used linear PEI

**Table 3.1** Molecular characterization of DNA/*b*PEI-25 K polyplexes (P) before and after addition of 7 portions of different free PEI chains

| Sample[a] | $M_w$/(g/mol)[b] | $<R_h>$ /nm[b] | $<\rho>$ /(g/cm$^3$)[b] | $\zeta_{potential}$/mV |
|---|---|---|---|---|
| P | $2.2 \times 10^7$ | 44 | 0.11 | $32.9 \pm 2.4$ |
| P + free *b*0.8 K | $2.2 \times 10^7$ | 43 | 0.11 | $32.0 \pm 2.9$ |
| P + free *b*2 K | $2.4 \times 10^7$ | 44 | 0.11 | $33.2 \pm 2.2$ |
| P + free *lin*2.5 K | $2.3 \times 10^7$ | 42 | 0.13 | $31.4 \pm 2.2$ |
| P + free *lin*25 K | $2.5 \times 10^7$ | 46 | 0.10 | $32.1 \pm 3.0$ |

[a] The total and final N:P ratio was kept at N:P = 10, identical for all the samples
[b] Determined by laser light scattering

chains with different lengths to complex with DNA in both salt-free water and salt-rich PBS and obtained similar results (not shown). Namely, (1) when N:P ~ 3 most of DNA are complexed with PEI; and (2) it is those uncomplexed PEI chains free in the solution mixture that greatly promote the gene transfection. On the basis of these results, we can use a combination of two different PEI chains: one to complex and condense DNA and the other with a different length as free chains. Before studying the gene transfection, we have known whether there is an exchange between bound chains inside the polyplexes and free chains in the solution mixture.

Theoretically, short PEI chains free in the solution mixture are not able to replace long PEI chains inside the polyplexes because the immobile of short chains leads to relatively more loss of the translational entropy so that the total entropy gain is less in comparison with the immobile of long chains. Table 3.1 shows that the addition of free PEI with different chain lengths has nearly no effect on physical parameters of the polyplexes made of long *b*PEI-25K chains. This result is further supported by the fluorescence assay, where long FITC-labeled *b*PEI-25K chains were used to complex DNA to form the polyplexes. The fluorescence intensity decreases due to the quench of fluorophores in proximity. Figure 3.3 shows that the addition of free PEI chains with different lengths has no effect on the fluorescence intensity of the DNA/FITC-*b*PEI-25K polyplexes (N:P = 3),

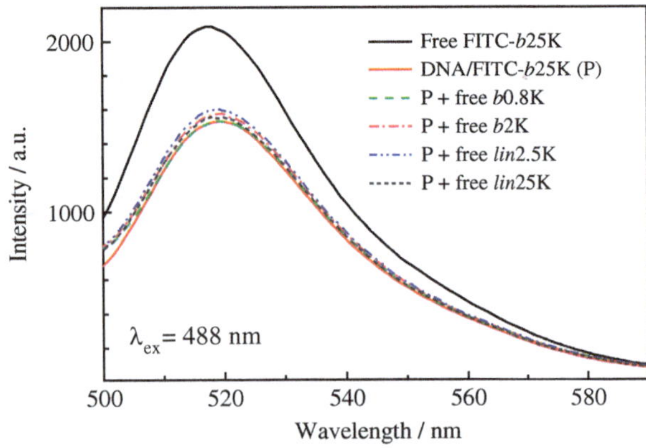


**Fig. 3.3** Effect of addition of 7 portions of different free PEI chains on the fluorescence intensity of DNA/FITC-*b*PEI-25K polyplexes (N:P = 3), where $C_{FITC\text{-}bPEI\text{-}25K}$ = 9.0 μg/mL, identical for all the tests

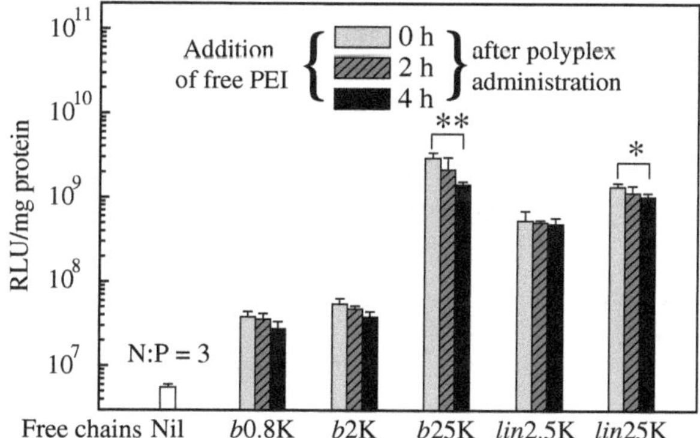


**Fig. 3.4** Effect of length and topology of free PEI chains on the gene transfection efficiency in 293T cells, where 7 portions of different PEIs were added at 0, 2 or 4 h after administrating the DNA/*b*PEI-25K polyplexes (N:P = 3). The total and final N:P ratio is 10, identical for all the tests. "Nil" means that no free PEI chains were added. ** indicates $p < 0.01$, and * indicates $p < 0.05$, $n = 3$, Student's *t* test

clearly indicating that there is no interchange between the *b*PEI-25K chains inside the polyplexes and short added branched or linear PEI chains free in the solution mixture. Otherwise, we would see an increase in the fluorescence intensity if there was some exchange between them.

Figure 3.4 shows the effect of the length of free PEI chains on the gene transfection when long *b*PEI-25K chains are used to complex and condense DNA, where $t = 0$ means a simultaneous addition of the DNA/*b*PEI-25K polyplexes (N:P = 3) and 7 portions of free PEI chains with different lengths; and $t > 0$ represents that free PEI chains were added after administrating the polyplexes. It is shown that the addition of free PEI chains essentially increases the gene transfection efficiency, irrespective of the chain length and topology (linear or branched). However, long chains, such as *b*PEI-25K and *lin*PEI-25K, free in the solution mixture are much more effective ($\sim 10^2$ fold) than their short counterparts. A combination of Figs. 3.1, 3.2, 3.3 and 3.4 shows that both short and long PEI chains are capable of condensing DNA completely at N:P $\sim$ 3 but free short PEI chains are less effective in enhancing the gene transfection, indicating that the chain length plays an important role in the PEI-mediated gene transfection. On the other hand, it is interesting to note that for long chains (25 K), the chain topology makes no big difference in the transfection efficiency, but for short chains ($\sim$2 K), linear PEI chains are $\sim$10 times more effective than branched ones. We will come back to this point later. Figure 3.4 also shows that the efficacy of long free chains only slightly decreases if they are added after the administration of the polyplexes, implying that the DNA/*b*PEI-25K polyplexes are fairly stable in the extra- and intra-cellular space up to 4 h.

Figure 3.5a shows the effect of free PEI concentration on the gene transfection, where a pair of short and long branched chains (*b*PEI-2K and *b*PEI-25K) is used.

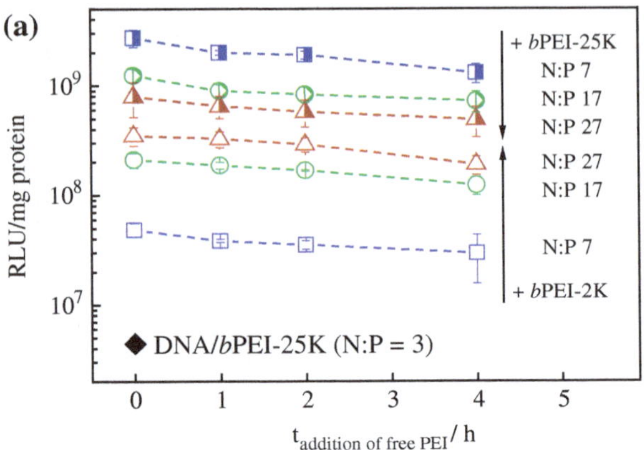


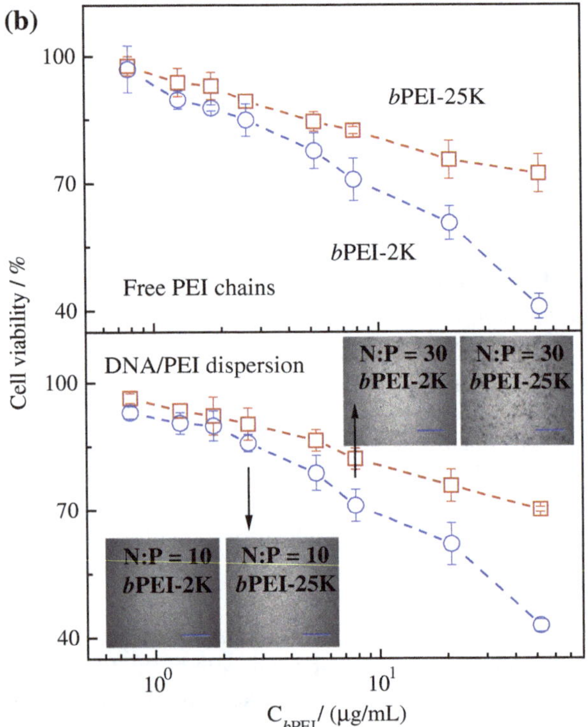


**Fig. 3.5** **a** Effect of free PEI concentration on gene transfection efficiency, where free short (*b*PEI-2 K) and long (*b*PEI-25 K) chains were respectively added at different times after the administration of DNA/*b*PEI-25K polyplexes (N:P = 3). **b** Concentration dependence of cell viability when PEI solution and DNA/PEI dispersion were respectively used, where MTT assay was used and $C_{bPEI}$ = 2.7 µg/mL corresponds to N:P = 10. *Insets* Microscopic images of 293T cells 6 h after administration of DNA/*b*PEI-2K and DNA/*b*PEI-25K dispersions, respectively. Scale bar: 200 µm

Taking N:P = 3 (no free chains) as a reference, we can see that for short chains, the transfection efficiency increases with the PEI concentration, from ~10-fold at N:P = 7 to ~80-fold at N:P = 27; but for long chains, the transfection efficiency slightly decreases as N:P increases, which can be attributed to the fact that long chains are much more cytotoxic than short ones, especially at higher concentrations, as shown in Fig. 3.5b. Quantitatively, it is worth noting that when

$C_{bPEI} > 0.8$ μg/mL, corresponding to N:P > 3, both the PEI solution and the DNA/PEI solution mixture exhibit a similar toxicity profile (Fig. 3.5b, $p > 0.05$, $n = 4$, Student's *t* test). Therefore, it is those free PEI chains, especially long ones, that lead to the cytotoxicity at high N:P ratios. On the other hand, we should note that N:P ~ 10 is typically used for the gene transfection, corresponding to $C_{bPEI} \sim 3$ μg/mL at which both *b*PEI-2K and *b*PEI-25K chains exhibit little cytotoxicity and the cell viability is well above 85 %. The pronounced cytotoxicity of long PEI chains is also reflected in their ability to alternate the cell morphology, which is attributed to the PEI-induced membrane disruption (also shown in Fig. 3.11) and apoptosis via mitochondrial pathway [46]. The inset of Fig. 5b reveals that after 6-h incubation with *b*PEI-25K at N:P = 30, some of the cell contours turn round and many cell debris appear, leading to a reduced cellular metabolic activity and a decrease of the transgene expression inside the cells.

Figure 3.6 shows the effect of the length and topology of the bound PEI chains on the gene transfection when long free PEI chains (*b*PEI-25K) are used. Very surprisingly, in spite that the polyplexes (N:P = 10, the filled symbols) made of different PEI chains have very different transfection efficiencies, the replacement of 7 portions of free chains with long ones (*b*PEI-25K) greatly enhances their efficiencies and reduces their differences when the polyplexes and long free chains are added at the same time (i.e., t = 0). To generalize such an issue, we repeated this experiment with the HeLa cells by using *lin*PEI-25K as free chains. The results confirm that upon a simultaneous addition of free *lin*PEI-25K chains, the polyplexes made of different PEIs display a similar high gene transfection efficiency, regardless of the length and topology of the bound PEI chains used (data not shown).

Figure 3.6 also shows that the gene transfection efficiency generally decreases as the delay time of adding free *b*PEI-25K chains increases. For the polyplexes made of long bound chains, the delay time has much less effect on the transfection efficiency, but for those made of short chains, especially *b*PEI-0.8K, the delayed addition of free chains significantly reduces their transfection efficiency. Likely,

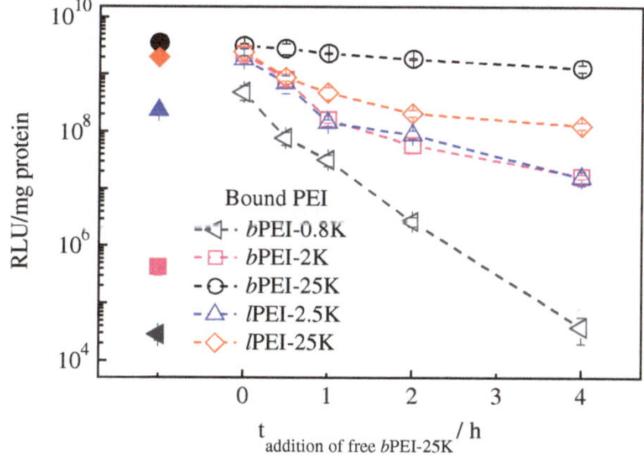


**Fig. 3.6** Effect of length and topology of bound PEI chains on the gene transfection efficiency in 293T cells, where 7 portions of free long chains (*b*PEI-25K) were added at different times after the polyplex administration (N.P = 3). The final and total N:P is 10, identical for all the tests. The filled symbols represent the transfection efficiency when the 7-portion free PEIs are identical to those bound chains

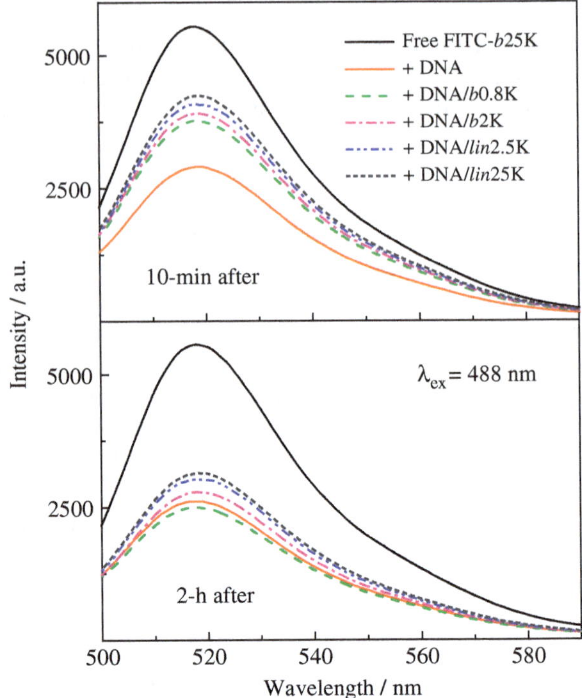


**Fig. 3.7** Effect of adding different PEI-mediated polyplexes (N:P = 3) on the fluorescence intensity of a PEI solution containing 7 portions of free FITC-labeled *b*PEI-25K chains, where $C_{\text{FITC-bPEI-25K}}$ is kept at 21 μg/mL

the addition of free *b*PEI-25K chains to the dispersion of the polyplexes made of *b*PEI-0.8K leads to a chain interchange so that long chains replace short ones, resulting in the formation of more stable DNA/*b*PEI-25K polyplexes. To confirm it, we respectively added naked DNA and different PEI-mediated polyplexes (N:P = 3) to a solution containing 7 portions of free FITC-*b*PEI-25K chains and monitored its fluorescence emission.

Figure 3.7 reveals that ~10 min after the polyplex administration, the fluorescence intensity of FITC-PEI in each case decreases and reaches equilibrium after 1.5–2.0 h, indicating the interchange between long *b*PEI-25K chains free in the solution and short branched or linear PEI chains inside the polyplexes. Namely, long branched PEI chains can replace most of short chains inside the polyplexes. In the gene transfection, when long *b*PEI-25K chains were added hours after the administration of the polyplexes made of short chains, the chain replacement should not be a problem because the polyplexes have already moved inside the cell. Previous studies showed that short PEI chains have a lower binding affinity to DNA than long ones. Therefore, the polyplexes made of short PEI chains would be more likely to release DNA in the presence of other long anionic chains, such as proteins or RNAs, in the intracellular space [47, 48], resulting in a lower transfection efficiency. Consequently, these less stable polyplexes made of short PEI chains were entrapped and digested inside the later endolysosomes before the addition of long free PEI chains. To exclude the influence of such a chain interchange, we mainly used long PEI chains (*b*PEI-25K) to condense DNA in the current study so that we can concentrate on the effect of length of free chains, mainly *b*PEI-2K and *b*PEI-25K, on the extra- and intra-cellular gene delivery. The real challenge is to elucidate how long free cationic PEI chains prevent the development of the later endolysosomes, help the escape of polyplexes from endosomes, and promote the gene transfection in the intracellular space, including nuclear localization and the finally de-assembly and transcription.

Since long free PEI chains significantly enhance the gene transfection, it is nature to ask whether they promote the cellular association and endocytosis in the extracellular space. To answer this question, we used the flow cytometry to monitor the cellular uptake kinetics of the DNA/PEI polyplexes in the presence of free PEI chains with different lengths. Generally, three characteristics can be extracted from one flow cytometry measurement, namely, (1) the fraction of cells containing Cy5-labeled DNA (Fig. 3.8a); (2) the median of the fluorescence intensity of Cy5-positive cell population ($F_{Cy5\text{-}DNA,\ Cy5+}$, Fig. 3.8b); and (3) the median of the fluorescence intensity of total cell population ($F_{Cy5\text{-}DNA,\ tot}$, Fig. 3.8c).

Figure 3.8a shows that in the presence of 7 portions of free short *b*PEI-2K or long *b*PEI-25K chains much more cells can internalize the polyplexes within 1 h. In each comparison group, the cellular uptake nearly ceases and the fluorescence intensity approaches a plateau 6 h after the incubation. The corresponding percentages of the cell internalization respectively reach 75, 95 and 65 %. On the other hand, Fig. 3.8b shows that $F_{Cy5\text{-}DNA,\ Cy5+}$ (an indication of the average amount of polyplexes inside each Cy5-positive cell) in the first 6-h incubation remains $\sim$ 2- or 10-fold higher when free *b*PEI-2K or *b*PEI-25K chains are added. Collectively, it is revealed that the addition of free PEI chains enhances the rate constant of the cellular uptake and long free PEI chains can make the uptake much faster than short ones. However, previous results demonstrated that the major play of free cationic PEI chains is mainly in the intracellular space [31]. The confocal live cell imaging revealed that most of the free PEI chains were co-localized together with the polyplexes in the endolysosomes [34], indicating that free chains might involve in the endolysosomal release. To have a more quantitative elucidation of such an involvement, we used a specific inhibitor for vacuolar ATPase proton pump, bafilomycin A1 (Baf-A1), to shut off the acidification of the early endosomes and prevent the development of the later endolysosomes [49, 50]. Naturally, the shut-off of proton pumps on the endolysosomes should also eliminate the so-called "proton-sponge" effect if any.

Figure 3.9a shows that the addition of baf-A1 reduces the gene transfection efficiency by a factor of $\sim$ 4, $\sim$ 6 and $\sim$ 16, respectively, for N:P = 3 without free chains, N:P = 10 with short *b*PEI-2K and long *b*PEI-25K free chains. Relatively, the reduced transfection efficiency with long *b*PEI-25K free chains (N:P = 10) is still $\sim$20 times higher than that without free chains (N:P = 3). It clearly indicates that even without the help of the so-called "proton pump" effect, long free PEI chains can still help the polyplexes to avoid the entrapment into the endolysosomes, presumably via the prevention of development of the later endolysomes or

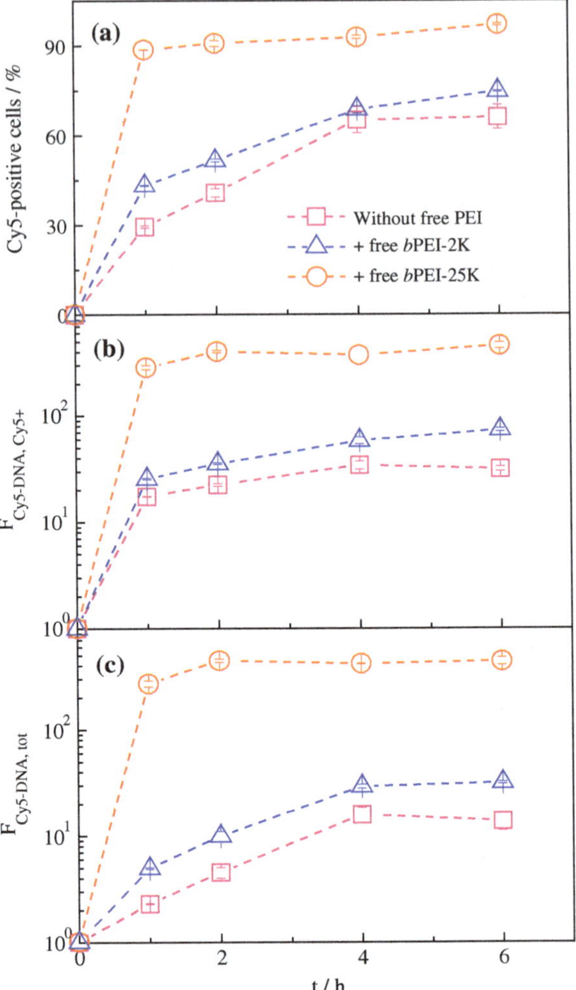


**Fig. 3.8** Flow cytometry study of effect of length of free PEI chains on cellular uptake of Cy5-DNA/*b*PEI-25K polyplexes, where 293T cells were used, the transfection was respectively carried without (N:P = 3) and with 7 portions of free *b*PEI-2 K or *b*PEI-25K chains (Final and total N:P = 10), and cellular uptake extent is expressed as **a** percentage of Cy5-positive cells; **b** median of Cy5-DNA fluorescence intensity of Cy5-positive cell population ($F_{Cy5\text{-}DNA,\ Cy5+}$); and **c** median of Cy5-DNA fluorescence intensity of total cell population ($F_{Cy5\text{-}DNA,\ tot}$)

the promotion of the escape from the early endosomes or both. Further, we tried to estimate when the polyplexes escape from the endolysosomes, respectively, without and with the addition of short *b*PEI-2K and long *b*PEI-25K free chains by adding baf-A1 at different times after the administration of the DNA/PEI polyplexes.

Figure 3.9b shows that ~6 h after the polyplexes administration, the addition of baf-A1 has nearly no effect on the gene transfection efficiency when N:P = 10 (with long *b*PEI-25K free chains), implying that most of the polyplexes are able to escape from the endolysosomes after ~6 h with the help of long free chains. Such an escaping time increases to ~10 h when short *b*PEI-2K free chains are used. Without free chains, the escaping time is longer than 12 h. To further elucidate the escape from the endolysosomes, we monitored the environmental pH near the

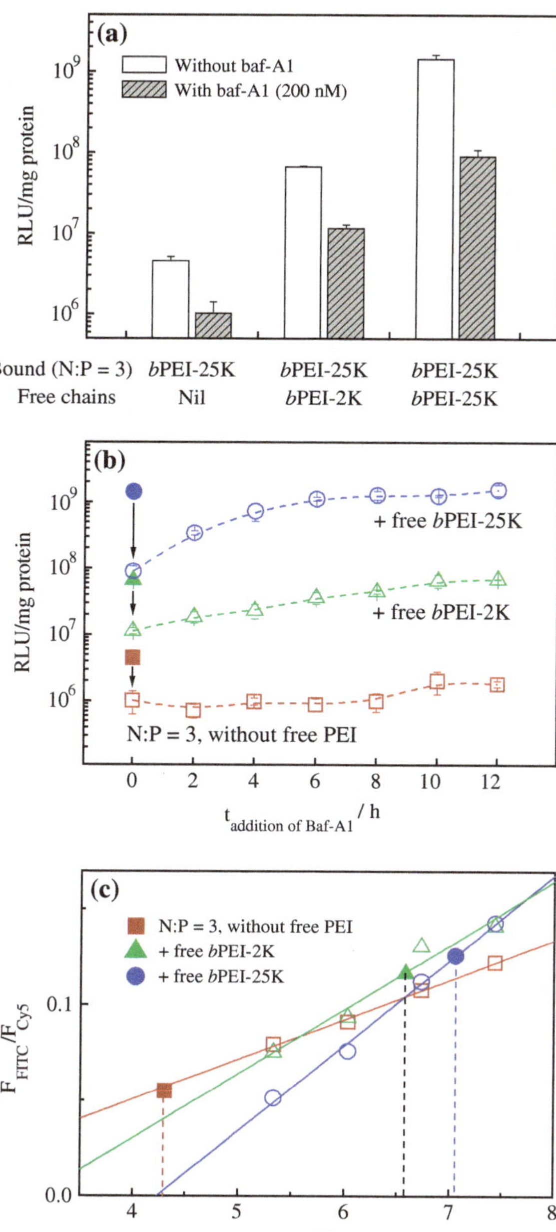


**Fig. 3.9** **a** Effect of simultaneous addition of bafilomycin A1 (baf-A1) on the transfection efficiency of DNA/*b*PEI-25K polyplexes (N:P = 3) without and with 7 portions of free *b*PEI-2K or *b*PEI-25K chains (total N:P = 10). **b** Time-dependent effect of baf-A1 treatment on the transfection efficiency of DNA/*b*PEI-25K polyplexes (N:P = 3) without and with 7 portions of free *b*PEI-2K or *b*PEI-25K chains (total N:P = 10), where filled symbols denote that no baf-A1 was added. **c** pH calibration curves (hollow symbols), where each symbol represents one study of DNA/*b*PEI-25K polyplexes (N:P = 3) without and with 7 portions of free PEI chains, and filled symbols show the intracellular pH near the polyplexes in each case, which were respectively obtained from the normalized relative fluorescence intensity via each calibration curve

polyplexes in the intracellular space, as shown in Fig. 3.9c. Our study reveals that 6 h after the polyplexes administration, without free PEI chains (N:P = 3), the intracellular pH near the polyplexes dramatically decreases from ~7.4 to ~4.3, clearly showing that the polyplexes alone are unable to escape from the acidic lysosomes. In contrast, with short *b*PEI-2K and long *b*PEI-25K free chains (N:P = 10), the intracellular pH near the polyplexes slightly decreases to ~6.6 and ~7.1, respectively, implying that the polyplexes have escaped into the cytoplasm or still reside inside the endosomes that are not matured into the later endolysosomes. Relatively speaking, short cationic PEI chains are not as effective as long ones. Our current results should convince readers that the "proton sponge" effect might not be related to the osmotic pressure and is only partially responsible for the safe trafficking of the polyplexes inside the cell. We have to consider that free PEI chains might involve other mechanisms in the development of the later endolysosomes, the escape of the polyplexes from the endosomes into the cytosol, and the nuclear localization, release and transcription.

As discussed before, the role of PEI in facilitating the endolysosomal release was previously attributed to its ability to increase the osmotic pressure inside the endolysosomes via the proton sponge effect [4, 29]or to destabilize the endolysosomal membrane via electrostatic interaction [15, 51]. Our previous results showed that a simple chemical coupling of few *b*PEI-2K chains via a linker can enhance the gene transfection by a factor of ~ $10^4$ times [8], which could not be explained by the increase of the internal osmotic pressure (the "proton sponge" effect) because we did not alter the chemical structure. Therefore, we have to further explore how different PEI chains interact with the phospholipid membrane by using narrowly distributed phospholipid vesicles ($\langle R_g \rangle = 48.3$ nm, $\langle R_h \rangle = 42.2$ nm, and $\zeta_{potential} = -38.6 \pm 0.9$ mV) and 293 T cells.

Thermodynamically, the attraction of cationic PEI chains to anionic vesicles is also entropy-driven due to the release of small counter ions (the gain of translational entropy). Therefore, long PEI chains lead to relatively less translational entropy lose than short ones so that the total gain in entropy is higher. Our study of the synthetic phospholipid vesicles reveals that long cationic *b*PEI-25K chains can reverse the charge of the vesicles at a much lower concentration (~ 2 μg/mL) than short *b*PEI-0.8K or *b*PEI-2K chains (Fig. 3.10). When applied to 293T cells, it is shown that long *b*PEI-25K chains are much more disruptive to the cellular

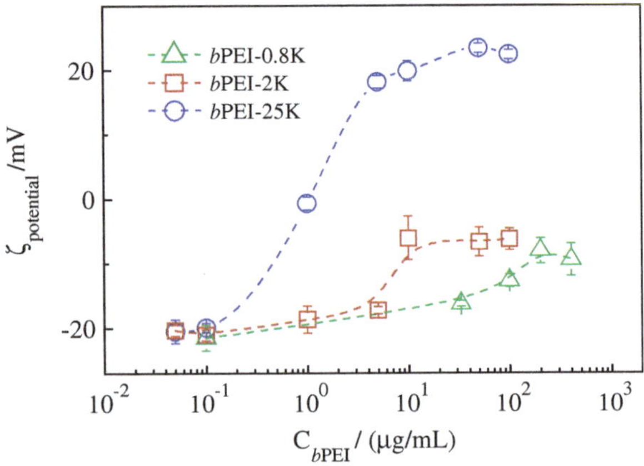


**Fig. 3.10** Zeta-potential ($\zeta_{potential}$) alteration profiles of the phospholipid vesicles upon the addition of different PEIs in 0.5 × PBS without NaCl and KCl

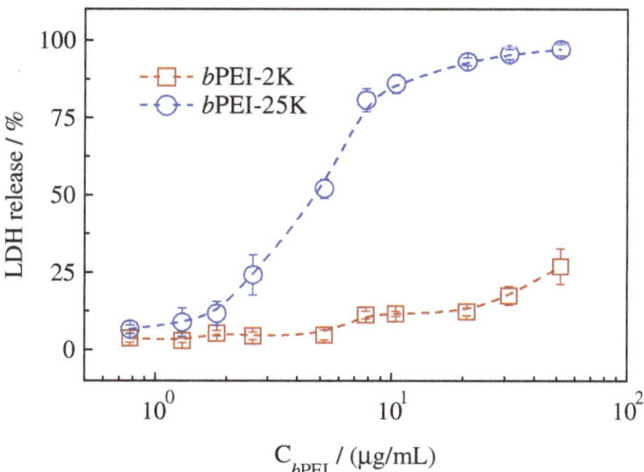


**Fig. 3.11** Concentration dependence of LDH release from 293T cells, 6 h after they are treated with *b*PEI-2K and *b*PEI-25K chains, where $C_{bPEI} = 2.7$ μg/mL corresponds to N:P = 10 used in an effective gene transfection

membrane than short *b*PEI-2K chains for a given polymer concentration, especially when $C_{bPEI} \geq 2.7$ μg/mL, corresponding to N:P ≥ 10 in our gene transfection experiments (Fig. 3.11). It seems that the destabilization and disruption of the cellular membrane by long free PEI chains is correlated to the release of the polyplexes from the endosomes in the intracellular space, and to some extent, the uptake of the polyplexes into the cells in the extracellular space.

Our previous and current results lead us to propose a hypothesis about why long free PEI chains are more effective in enhancing the gene transfection as follows. When long PEI chains are added into the cell culture medium, they are able to quickly penetrate different membranes of the cell and across the cytosol all the way into nucleus within less than one hour (data not shown). Some of them are embedded inside the membranes. Typical phospholipid bi-layer membranes with two anionic surfaces have a thickness of 5–6 nm. It is those embedded cationic chains that weaken/destabilize the anionic membrane via electrostatic interaction to facilitate the escape of polyplexes from the endosomes. More importantly, in order to differentiate the endosomes entrapping foreign subjects from those vesicles generated by different organelles inside the cell, lysosomes normally use the membrane proteins attached to the inner surface of the cell membrane as a signal. The embedment of long cationic PEI chains on the cell membrane might shield those signal proteins so that the fusion between endosomes and lysosomes and the development of the later endolysosomes are prevented or delayed. Using this hypothesis, we can explain many of those observed differences in the gene transfection efficiency, such as (1) why short free PEI chains are less effective because they are too short to shield the membrane proteins (e.g. *b*PEI-0.8K ~ 4 nm and *b*PEI-2K ~ 6 nm); (2) why *lin*PEI-2.5K is more effective than *b*PEI-2K because linear PEI-2.5K chains (~18 nm if stretched) are much longer than their branched counterparts even they have an identical molar mass; (3) why the coupling of 3–4 short *b*PEI-2K chains into a long one (~18 nm) can enhance the transfection efficiency by a factor of $\sim 10^4$ times; and (4) why *lin*PEI-25K and *b*PEI-25K chains have a similar transfection efficiency even though linear ones have a much longer contour length because linear chains are coiled in solution so that they have a similar size as the branched ones. It should be noted that the longer the cationic chains, the more cytotoxic they become. Therefore, there is a dedicate balance between the gene transfection efficiency and the cytotoxicity. Our hypothesis essentially suggests that one should choose cationic chains with a proper length ( ~ 15–20 nm if stretched) so that they are able to (1) insert themselves into the cell membrane via electrostatic interaction, and at the same time, (2) stick out their end(s) to shield the membrane proteins attached on the inner surface of the cell membrane. Such a strategy was previously exploited but not established on the above hypothesis. For example, Mahato et al. [52] have shown that attaching one cholesterol to each short *b*PEI-1.8K chain greatly boosted its transfection efficiency, while the modification on *b*PEI-10K had no such enhancement in the gene transfection, presumably because *b*PEI-10K is long enough to insert into the membranes. Alternatively, Uludag et al. [53] modified *b*PEI-2K with a set of aliphatic lipids with different lengths and found that attaching aliphatic lipid to *b*PEI-2K can transform the ineffective *b*PEI-2K into an effective non-viral vector. Notably, linoleic acid (LA, $C_{17}H_{31}CO$–) and palmitic acid (PA, $C_{15}H_{31}CO$–) substituted PEI derivatives led to a much higher gene transfection efficiency than caprylic acid (CA, $C_7H_{15}CO$–) substituted ones (lipid: PEI molar ratio ~ 1), comparable to the potent *b*PEI-25K but much less cytotoxic.

## 3.4 Conclusion

A combination of laser light scattering (LLS) and gel electrophoresis results confirm that most of anionic DNA chains are complexed and condensed by cationic PEI chains when the molar ratio of nitrogen from PEI to phosphate from DNA (N:P) reaches ~ 3, irrespective of the chain length and solvent (pure water or PBS). In the solution mixture (N:P > 3), there exists two kinds of PEI chains: bound to DNA and free in the solution mixture. The bound chains mainly provide the charge neutrality, resulting in the condensation and collapse of long plasmid DNA chains into $10^2$-nm particles. It is those free PEI chains, rather than other physical properties of the DNA/PEI polyplexes, that play a vital role in promoting the gene transfection by a factor of $10^2$–$10^3$ times, depending on the chain length. Our results reveal that long *b*PEI-25K free chains are ~ $10^2$ more effective than their short counterparts (*b*PEI-0.8K and *b*PEI-2K). Long free cationic chains can significantly increase the uptake rate constant of the polyplexes, presumably due to their ability to disrupt the anionic cell membrane via electrostatic interaction. However, the main contribution of these long free PEI chains is in the intracellular space; namely, they prevent the entrapment of the DNA/PEI polyplexes into the endolysosomes. We can conclude that the increase of the osmotic pressure inside the later endolysosomes due to the "proton sponge" effect plays a certain role, but not dominant, in promoting the gene transfection. The current study leads to a hypothesis as follows. It is those free cationic PEI chains with a proper length

($\sim$15–20 nm) embedded inside the anionic cell membrane via electrostatic interaction that are able to (1) destabilize/weaken the endosome membrane (the "proton sponge" effect might help here) and promote the escape of the polyplexes entrapped inside; and (2) prevent the fusion between the endosomes and lysosomes so that further development of the later endolysosomes is hindered. This is because the stick-out cationic chain end(s) effectively shield those "signaling" anionic proteins embedded on the inner surface of the cell membrane (i.e., the outer surface of the endosomes). Note that the development of non-viral vectors is an extremely complicate and multi-dimensional problem. Therefore, some fundamental understanding is not only needed but also a must. We should do a hypothesis-driven research instead of a "lottery-like" trial-and-error approach. The huge amount of literature (>$10^5$ publications) generated in the past 3–4 decades have already taught us a great lesson. On the basis of our hypothesis, we will modify short PEI chains ($\sim$5 nm) with a proper hydrophobic molecule (chain segment) ($\sim$2–5 nm in size) so that they can effectively stick inside the cell membrane and expose short PEI chains on its surface; and use them as the 7 portions of free cationic chains to promote the gene transfection. In this way, we hope to be able to solve the catch-22 "transfection efficiency"-versus-"cytotoxicity" problem because it has been known that short PEI chains are nearly not cytotoxic in a normal concentration range used for the gene transfection. Further, we will develop some methods to pack them together with the polyplexes so that they can be safely delivered to the targeted cells or organs in in vivo experiments.

## References

1. Pack, D. W., Hoffman, A. S., Pun, S., & Stayton, P. S. (2005). Design and development of polymers for gene delivery. *Nature Reviews Drug Discovery, 4*, 581–593.
2. Li, S. D., & Huang, L. (2007). Non-viral is superior to viral gene delivery. *Journal of Controlled Release, 123*, 181–183.
3. Mintzer, M. A., & Simanek, E. E. (2009). Nonviral vectors for gene delivery. *Chemical Reviews, 109*, 259–302.
4. Boussif, O., Lezoualch, F., Zanta, M. A., Mergny, M. D., Scherman, D., Demeneix, B., et al. (1995). A versatile vector for gene and oligonucleotide transfer into cells in culture and in vivo-polyethylenimine. *Proceedings of the National Academy of Sciences USA, 92*, 7297–7301.
5. Lungwitz, U., Breunig, M., Blunk, T., & Gopferich, A. (2005). Polyethylenimine-based non-viral gene delivery systems. *European Journal of Pharmaceutics and Biopharmaceutics, 60*, 247–266.
6. Neu, M., Fischer, D., & Kissel, T. (2005). Recent advances in rational gene transfer vector design based on poly(ethylene imine) and its derivatives. *Journal of Gene Medicine, 7*, 992–1009.
7. Breunig, M., Lungwitz, U., Liebl, R., & Goepferich, A. (2007). Breaking up the correlation between efficacy and toxicity for nonviral gene delivery. *Proceedings of the National Academy of Sciences USA, 104*, 14454–14459.
8. Deng, R., Yue, Y., Jin, F., Chen, Y. C., Kung, H. F., Lin, M. C. M., et al. (2009). Revisit the complexation of PEI and DNA—how to make low cytotoxic and highly efficient PEI gene

transfection non-viral vectors with a controllable chain length and structure? *Journal of Controlled Release, 140*, 40–46.
9. Ogris, M., Brunner, S., Schuller, S., Kircheis, R., & Wagner, E. (1999). PEGylated DNA/transferrin–PEI complexes: reduced interaction with blood components, extended circulation in blood and potential for systemic gene delivery. *Gene Therapy, 6*, 595–605.
10. Cheng, H., Zhu, J. L., Zeng, X., Jing, Y., Zhang, X. Z., & Zhuo, R. X. (2009). Targeted gene delivery mediated by folate-polyethylenimine-block-poly(ethylene glycol) with receptor selectivity. *Bioconjugate Chemistry, 20*, 481–487.
11. Zanta, M. A., Boussif, O., Adib, A., & Behr, J. P. (1997). In vitro gene delivery to hepatocytes with galactosylated polyethylenimine. *Bioconjugate Chemistry, 8*, 839–844.
12. Diebold, S. S., Kursa, P., Wagner, E., Cotten, M., & Zenke, M. (1999). Mannose polyethylenimine conjugates for targeted DNA delivery into dendritic cells. *Journal of Biological Chemistry, 274*, 19087–19094.
13. Pollard, H., Remy, J. S., Loussouarn, G., Demolombe, S., Behr, J. P., & Escande, D. (1998). Polyethylenimine but not cationic lipids promotes transgene delivery to the nucleus in mammalian cells. *Journal of Biological Chemistry, 273*, 7507–7511.
14. Godbey, W. T., Wu, K. K., & Mikos, A. G. (1999). Tracking the intracellular path of poly(ethylenimine)/DNA complexes for gene delivery. *Proceedings of the National Academy Sciences USA, 96*, 5177–5181.
15. Bieber, T., Meissner, W., Kostin, S., Niemann, A., & Elsasser, H. P. (2002). Intracellular route and transcriptional competence of polyethylenimine–DNA complexes. *Journal of Controlled Release, 82*, 441–454.
16. Thomas, M., & Klibanov, A. M. (2002). Enhancing polyethylenimine's delivery of plasmid DNA into mammalian cells. *Proceedings of the National Academy Sciences USA, 99*, 14640–14645.
17. Sonawane, N. D., Szoka, F. C., & Verkman, A. S. (2003). Chloride accumulation and swelling in endosomes enhances DNA transfer by polyamine–DNA polyplexes. *Journal of Biological Chemistry, 278*, 44826–44831.
18. Suh, J., Wirtz, D., & Hanes, J. (2003). Efficient active transport of gene nanocarriers to the cell nucleus. *Proceedings of the National Academy Sciences USA, 100*, 3878–3882.
19. Akinc, A., Thomas, M., Klibanov, A. M., & Langer, R. (2005). Exploring polyethylenimine mediated DNA transfection and the proton sponge hypothesis. *Journal of Gene Medicine, 7*, 657–663.
20. Kulkarni, R. P., Wu, D. D., Davis, M. E., & Fraser, S. E. (2005). Quantitating intracellular transport of polyplexes by spatio-temporal image correlation spectroscopy. *Proceedings of the National Academy Sciences USA, 102*, 7523–7528.
21. de Bruin, K. G., Fella, C., Ogris, M., Wagner, E., Ruthardt, N., & Brauchle, C. (2008). Dynamics of photoinduced endosomal release of polyplexes. *Journal of Controlled Release, 130*, 175–182.
22. Gabrielson, N. P., & Pack, D. W. (2009). Efficient polyethylenimine-mediated gene delivery proceeds via a caveolar pathway in HeLa cells. *Journal of Controlled Release, 136*, 54–61.
23. Won, Y.-Y., Sharma, R., & Konieczny, S. F. (2009). Missing pieces in understanding the intracellular trafficking of polycation/DNA complexes. *Journal of Controlled Release, 139*, 88–93.
24. Lee, H., Kim, I. K., & Park, T. G. (2010). Intracellular trafficking and unpacking of siRNA/quantum Dot–PEI complexes modified with and without cell penetrating peptide:confocal and flow cytometric FRET analysis. *Bioconjugate Chemistry, 21*, 289–295.
25. Wagner, E., Cotten, M., Foisner, R., & Birnstiel, M. L. (1991). Transferrin polycation DNA complexes—the effect of polycations on the structure of the complex and DNA delivery to cells. *Proceedings of the National Academy Sciences USA, 88*, 4255–4259.
26. Godbey, W. T., Wu, K. K., & Mikos, A. G. (1999). Size matters: molecular weight affects the efficiency of poly(ethylenimine) as a gene delivery vehicle. *Journal of Biomedical Materials Research, 45*, 268–275.

27. Kunath, K., von Harpe, A., Fischer, D., Peterson, H., Bickel, U., Voigt, K., et al. (2003). Lowmolecular-weight polyethylenimine as a non-viral vector for DNA delivery:comparison of physicochemical properties, transfection efficiency and in vivo distribution with high-molecular-weight polyethylenimine. *Journal of Controlled Release, 89*, 113–125.
28. Honore, I., Grosse, S., Frison, N., Favatier, F., Monsigny, M., & Fajac, I. (2005). Transcription of plasmid DNA: influence of plasmid DNA/polyethylenimine complex formation. *Journal of Controlled Release, 107*, 537–546.
29. Behr, J. P. (1997). The proton sponge: a trick to enter cells the viruses did not exploit. *Chimia, 51*, 34–36.
30. Forrest, M. L., Meister, G. E., Koerber, J. T., & Pack, D. W. (2004). Partial acetylation of polyethylenimine enhances in vitro gene delivery. *Pharmaceutical Research, 21*, 365–371.
31. Yue, Y., Jin, F., Deng, R., Cai, J., Chen, Y., Lin, M. C. M., et al. (2011). Revisit complexation between DNA and polyethylenimine—effect of uncomplexed chains free in the solution mixture on gene transfection. *Journal of Controlled Release, 155*, 67–76.
32. Clamme, J. P., Azoulay, J., & Mely, Y. (2003). Monitoring of the formation and dissociation of polyethylenimine/DNA complexes by two photon fluorescence correlation spectroscopy. *Biophysical Journal, 84*, 1960–1968.
33. Clamme, J. P., Krishnamoorthy, G., & Mely, Y. (2003). Intracellular dynamics of the gene delivery vehicle polyethylenimine during transfection: investigation by two photon fluorescence correlation spectroscopy. *Biochimica et Biophysica Acta, Biomembranes, 1617*, 52–61.
34. Boeckle, S., von Gersdorff, K., van der Piepen, S., Culmsee, C., Wagner, E., & Ogris, M. (2004). Purification of polyethylenimine polyplexes highlights the role of free polycations in gene transfer. *Journal of Gene Medicine, 6*, 1102–1111.
35. Fahrmeir, J., Gunther, M., Tietze, N., Wagner, E., & Ogris, M. (2007). Electrophoretic purification of tumor-targeted polyethylenimine-based polyplexes reduces toxic side effects in vivo. *Journal of Controlled Release, 122*, 236–245.
36. Saul, J. M., Wang, C. H. K., Ng, C. P., & Pun, S. H. (2008). Multilayer nanocomplexes of polymer and DNA exhibit enhanced gene delivery. *Advanced Materials, 20*, 19–25.
37. Chu, B. (1991). *Laser light scattering* (2nd ed.). New York: Academic Press.
38. Wu, C., & Xia, K. Q. (1994). *Review of Scientific Instruments, 65*, 587–590.
39. Berne, B., & Pecora, R. (1976). *Dynamic light scattering*. New York: Plenum Press.
40. Hunter, R. J. (2000). *Foundations of colloid science* (2nd ed.). Oxford: Oxford Press.
41. Forrest, M. L., & Pack, D. W. (2002). On the kinetics of polyplex endocytic trafficking:implications for gene delivery vector design. *Molecular Therapy, 6*, 57–66.
42. Breunig, M., Lungwitz, U., Liebl, R., Fontanari, C., Klar, J., Kurtz, A., et al. (2005). Gene delivery with low molecular weight linear polyethytenimines. *Journal of Gene Medicine, 7*, 1287–1298.
43. Akinc, A. (2002). Measuring the pH environment of DNA delivered using nonviral vectors: implications for lysosomal trafficking. *Biotechnology Bioengineering, 78*, 503–508.
44. Peng, S. F., & Wu, C. (1999). Light scattering study of the formation and structure of partially hydrolyzed poly(acrylamide)/calcium(II) complexes. *Macromolecules, 32*, 585–589.
45. Zhang, G. Z., & Wu, C. (2006). Folding and formation of mesoglobules in dilute copolymer solutions. *Advances in Polymer Science, 195*, 101–176.
46. Rejman, J., Bragonzi, A., & Conese, M. (2005). A two-stage poly(ethylenimine)-mediated cytotoxicity: implications for gene transfer/therapy. *Molecular Therapy, 12*, 468–474.
47. Itaka, K., Harada, A., Yamasaki, Y., Nakamura, K., Kawaguchi, H., & Kataoka, K. (2004). In situ single cell observation by fluorescence resonance energy transfer reveals fastintra-cytoplasmic delivery and easy release of plasmid DNA complexed with linear polyethylenimine. *Journal Gene Medicine, 6*, 76–84.
48. Bertschinger, M., Backliwal, G., Schertenleib, A., Jordan, M., Hacker, D. L., & Wurm, F. M. (2006). Disassembly of polyethylenimine–DNA particles in vitro: implications for polyethylenimine-mediated DNA delivery. *Journal of Controlled Release, 116*, 96–104.

49. Bowman, E. J., Siebers, A., & Altendorf, K. (1988). Bafilomycins—a class of inhibitors of membrane Atpases from microorganisms, animal-cells, and plant-cells. *Proceedings of the National Academy of Sciences USA, 85*, 7972–7976.
50. Crider, B. P., Xie, X. S., & Stone, D. K. (1994). Bafilomycin inhibits proton flow-through the H+ channel of vacuolar proton pumps. *Journal of Biological Chemistry, 269*, 17379–17381.
51. Pichon, C., Roufai, M. B., Monsigny, M., & Midoux, P. (2000). Histidylated oligolysines increase the transmembrane passage and the biological activity of antisense oligonucleotides. *Nucleic Acids Research, 28*, 504–512.
52. Wang, D. A., Narang, A. S., Kotb, M., Gaber, A. O., Miller, D. D., Kim, S. W., et al. (2002). Novel branched poly(ethylenimine)-cholesterol water-soluble lipopolymers for gene delivery. *Biomacromolecules, 3*, 1197–1207.
53. Neamnark, A., Suwantong, O., Bahadur, K. C. R., Hsu, C. Y. M., Supaphol, P., & Uludag, H. (2009). Aliphatic lipid substitution on 2 kDa polyethylenimine improves plasmid delivery and transgene expression. *Molecular Pharmaceutics, 6*, 1798–1815.

# Chapter 4
# Quantitative Comparison of Endocytosis and Intracellular Trafficking of DNA/Polymer Complexes in the Absence/Presence of Free Polycationic Chains

## 4.1 Introduction

The development of smart and multi-functional non-viral polymeric devices (NVPD) for nucleic acid delivery has captured tremendous attention over the past few decades [1–3]. Particularly, cationic polymers are perceived as promising candidates due to their safety profiles and relative ease to incorporate desired functional moieties [4, 5]. Currently, polyethylenimine (PEI) is among the most intensively studied cationic polymeric vehicles and hitherto mediates nearly the highest *in vitro* transfection efficiency in the absence of any exogenous endosomolytic agent [6–8]. Great efforts on synthetic chemistry has been made to construct a multi-potent PEI-based NVPD, including grafting poly(ethylene glycol) (PEG) on PEI to improve its *in vivo* circulation stability [9, 10], conjugating desired ligands on its surface to enhance its cell-targeting ability [10–12], introducing proper biodegradable linkers to its backbone to facilitate the intracellular unloading of its therapeutic cargo [13–18], attaching nuclear import moieties to it to stimulate the nuclear entry [19]. However, in most of the cases, their transfection efficacy remains orders of magnitude lower than the recombinant virus, especially in the *in vivo* animal tests and clinical trials [20]. This is, at least partially, due to the lack of some fundamental understanding on how DNA is delivered into the cell nucleus; namely, a clear and detailed pathway for the intracellular trafficking of the DNA/polymer complexes [20–22].

Recently, the complexation between the anionic DNA and cationic PEI chains in a physiological environment has been revisited by using a combination of laser light scattering and gel electrophoresis [23, 24]. The results clearly confirmed that nearly all the DNA chains are complexed with PEI to form small polyplexes ($\sim 10^2$ nm) when N:P $\sim$ 3, irrespective of the PEI chain length and topology [24]. However, a high *in vitro* gene transfection efficiency is only achieved when N:P $\geq$ 10. Putting these two facts together, it has been concluded that (1) each solution mixture with a higher N:P ratio actually contains two kinds of cationic chains: bound to DNA and free in the solution ($\sim$70 %), as schematically shown in Fig. 4.1a; and (2) it is the free cationic PEI chains, especially those long ones,

Y. Yanan, *How Free Cationic Polymer Chains Promote Gene Transfection*, Springer Theses, DOI: 10.1007/978-3-319-00336-8_4,

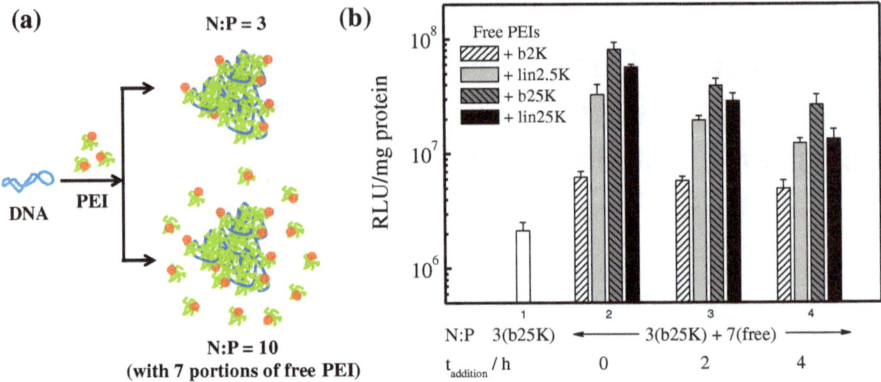


**Fig. 4.1** **a** Schematic of complexation between anionic DNA and cationic bPEI chains in solution. **b** Effect of length and topology of free PEI chains on the gene transfection efficiency in HepG2 cells, where 7 portions of different free PEI chains were added at 0, 2 or 4 h post-administration of DNA/*b*PEI-25K polyplexes (N:P = 3, denoted as "3(b25K)"). The total and final N:P ratio is 10, identical for all the tests

that actually promote the gene transfection no matter whether they exist (are applied) hours before or after the administration of the polyplexes (N:P = 3). These findings were further delineated by different polymers (PLL and PDMAEMA) and cell lines (293T, HepG2, HeLa and COS-7). Note that such an effect of free polycationic chains on *in vitro* gene transfection was previously reported but the detailed intracellular mechanism remains elusive [25–29].

Currently, it remains a challenge to elucidate how those free polycationic chains facilitate the intracellular trafficking of the polyplexes since direct observation of the trafficking events among different organelles is rather difficult. In clathrin-mediated (CME) pathway, much evidence has been accumulated so far to indicate that the well-accepted "proton sponge" effect plays a certain, but not dominant and decisive, role, in preventing the entrapment of polyplexes into the late endosomes/lysosomes [20, 30–39]. Our previous study also found that even after the complete removal of the "proton sponge" effect, long free cationic PEI chains still help to avoid the degradation of the DNA payload inside lysosomes [24]. In addition to the well-documented CME pathway, it is gradually realized that the caveolae-mediated pathway, a clathrin-independent route, also has a significant impact on the safe intracellular trafficking of polyplexes [34–36].However, real-time observation of these intracellular processes with high spatio-temporal resolution, and quantitative profiles on the polyplexes subcellular distribution are still largely lacking [37–39].

In the present study, we designed and utilized a stringent of quantitative methods, including real-time confocal three-dimensionally integrated quantification and flow cytometry, to address how free PEI chains with a proper length/

topology promote the endocytosis and the subsequent intracellular trafficking of polyplexes, particularly in the CME pathway. Moreover, we also investigated the effect of caveolae-mediated pathway on the intracellular fate of polyplexes, and more importantly, explored the possible role of long free PEI chains in this intracellular route. A possible mechanism on how free polycationic chains with a sufficient length (15–20 nm) facilitate polyplex trafficking in these two pathways is speculated and partially validated.

## 4.2 Materials and Methods

### *4.2.1 Materials and Cell Lines*

Two branched PEIs ($M_w$ = 2,000 and 25,000 g/mol, denoted as *b*PEI-2K and *b*PEI-25K) and two linear PEIs ($M_w$ = 2,500 and 25,000 g/mol, denoted as *lin*-PEI-2.5K and *lin*PEI-25K) were purchased from Sigma-Aldrich and Polysciences, respectively, and used without further purification. FITC-labeled *b*PEI-25K was prepared as previously described [23]. Initial plasmid DNA pGL3-control vector (5,256 bp) encoding modified firefly luciferase was purchased from Promega (USA), while a large amount of it was prepared by using a Qiagen Plasmid Maxi Kit (Qiagen, Germany). 3-(4,5-dimethylthiazol-2-yl)-2,5-diphenyltetrazolium bromide (MTT), chlorpromazine, filipin III and genistein were purchased from Sigma-Aldrich (Deutschland). CellLight-lysosomes GFP and LysoTracker Red DND-99 were purchased from Invitrogen (USA). The Bright-Glo luminescence assay and CytoTox 96 non-radioactive cytotoxicity assay kits were purchased from Promega (USA). The fetal bovine serum (FBS), Dulbecco's modified Eagle's medium (DMEM) and penicillin–streptomycin were products of GIBCO (USA). HepG2 and HeLa cells were grown at 37 °C, 5 % $CO_2$ in DMEM supplemented with 10 % FBS, penicillin at 100 units/mL and streptomycin at 100 μg/mL.

### *4.2.2 Formation of DNA/PEI Polyplexes*

The plasmid DNA (*p*DNA) was complexed with different PEIs in either distilled water or phosphate buffered saline (PBS) to form the DNA/PEI polyplexes as described [23].Briefly, different amounts of the PEI solution ($C = 10^2$–$10^3$ μg/mL) were added dropwise into a dilute DNA solution ($C$ = 14.5 μg/mL), resulting in different molar ratios of nitrogen from PEI to phosphate from DNA (N:P). Each resultant dispersion was incubated for 5 min at room temperature before being added to the cell culture medium.

### 4.2.3 In Vitro Gene Transfection

The *in vitro* gene transfection efficiency was quantified by using the luciferase transfection assays, in which plasmid pGL3 was used as an exogenous reporter gene. HepG2 and HeLa cells were seeded in a 48-well plate at an initial density of 30,000 and 40,000 cells per well, respectively, to ensure the cell confluence reaches ~80 % at the time of the gene transfection. After 24 h, each DNA/PEI dispersion with a desired N:P ratio was further diluted in serum-free medium and then administered to the cells at a final concentration of 0.4 μg DNA per well. For endocytosis inhibition experiments, HeLa cells were pretreated with one of the following inhibitors chlorpromazine (21 μM), filipin (3.8 μM) and genistein (100 μM) in serum-free DMEM for 1.5 h prior to the polyplex administration. The cells were further incubated with the polyplexes for 4 h at 37 °C before the medium was replaced with DMEM supplemented with 10 % FBS (300 μL/well). Using a GloMax 96 microplate luminometer (Promega, USA) and the Bio-Rad protein assay reagent, we respectively determined the transgene expression level and the corresponding protein concentration in each well 36 h after the polyplex administration. The gene transfection efficiency is expressed as a relative luminescence unit (RLU) per cellular protein (mean ± SD of quadruple).

### 4.2.4 MTT Assay

HepG2 cells were seeded in a 96-well plate at an initial density of 10,000 cells per well. After 24 h, various free PEI chains were added to the cells at different chosen concentrations. The treated cells were incubated in a humidified environment with 5 % $CO_2$ at 37 °C for 48 h. The MTT reagent (in 20 μL PBS, 5 mg/mL) was then added to each well. The cells were further incubated for 4 h at 37 °C. The medium in each well was then removed and replaced by 100 μL of DMSO. The plate was gently agitated for 15 min before the absorbance (*A*) at 490 nm was recorded by a microplate reader (Bio-rad, USA). The cell viability (*x*) was calculated by $x = (A_{treated}/A_{control}) \times 100\ \%$, where $A_{treated}$ and $A_{control}$ are the absorbance of the cells cultured with PEI and fresh culture medium, respectively. Each experiment condition was done in quadruplicate. The data was shown as the mean value plus a standard deviation (±SD).

### 4.2.5 Lactate Dehydrogenase (LDH) Membrane Integrity Assay

HepG2 cells were seeded in a 96-well plate at an initial density of 10,000 cells per well. After 24 h, cells were treated with various PEIs at different desired concentrations. The treated cells were incubated in a humidified environment with 5 % $CO_2$ at 37 °C for 6 h. To determine the maximum LDH release, 10 × lysis buffer was added to the control group 2 h prior to the usage of a CytoTox 96 Non-radioactive cytotoxicity assay kit. The degree of LDH Release, defined as $(A_{treated}-A_{control})/(A_{maximum}-A_{control}) \times 100\ \%$, represents the membrane disruption induced by different PEIs, where $A_{treated}$ and $A_{control}$ are the absorbance values of the cells cultured with and without PEI, and $A_{maximum}$ is the absorbance value of the cells in the maximum LDH release group. Each experiment condition was done in quadruplicate. Data was shown as the mean value plus a standard deviation (±SD).

### *4.2.6 Confocal Laser Scanning Microscopy*

Plasmid pGL3 was covalently labeled with the fluorophore Cy3 using a Label IT nucleic acid labeling kit (Mirus, Madison, WI) as the manufacturer's instructions. A total of 120,000 HepG2 cells were seeded in a $\mu$-Dish35 mm, high (ibidi GmbH, Germany). After cell adhesion (~4 h), lysosomes were labeled with CellLight lysosomes-GFP agent (particle per cell is 30). After ~24 h, Cy3-DNA/*b*PEI-25K polyplexes without and with different free PEI chains were respectively added to the cells ($C_{DNA}$ = 3.2 μg/mL). Live cell imaging was performed using a Nikon C1si CLSM equipped with a standard fluorescence detector (Nikon, Japan) and an INU stage-top incubator (Tokai Hit, Japan). Lysosomes-GFP and Cy3-DNA were excited at 488 and 543 nm, respectively; and the corresponding emissions were detected at 515/30 and 605/75 nm, respectively. Methodology to quantify the cellular uptake amount of polyplexes and the fraction of polyplexes entrapped inside lysosomes were described in detail in Results and discussion. All the images were analyzed using the Nikon EZ-C1 software.

### *4.2.7 Measurement of Intracellular pH Around Polyplexes*

Plasmid pGL3 was covalently double-labeled with pH-sensitive FITC and pH-insensitive Cy5 using a Label IT nucleic acid labeling kit (Mirus, Madison, WI). HepG2 cells were seeded in a 6-well plate at an initial density of 240,000 cells per well. After 24 h, double-labeled pGL3/*b*PEI-25K polyplexes (N:P = 3) without and with 7 portions of free *b*PEI-25K chains (N:P = 10) were respectively added to the HepG2 cells in serum-free DMEM. After 6-h incubation at 37 °C, HepG2 cells were harvested as described before [24] and pelleted into six separate eppendorf vials. The cells in two vials were resuspended with PBS containing 2 % FBS, whereas the cells in the rest four vials were respectively resuspended with four intracellular pH clamping buffers (pH = 5.33, 6.03, 6.73 and 7.43) [40].The cells in each vial were further washed by pelleting and resuspending in an appropriate and corresponding buffer. The FITC/Cy5 fluorescence ratio of each sample was measured using the flow cytometer. The fluorophores FITC and Cy5 were excited at 488 nm and 635 nm, respectively. The corresponding emissions were detected at 525/10 and 675/15 nm, respectively. The FITC/Cy5 fluorescence ratio was calculated using the median values of the FITC and Cy5 fluorescence intensities of the total cell population (10,000 cells). The four intracellular pH clamped samples generate a linear pH calibration, from which the FITC/Cy5 ratio of each of the other two duplicate samples corresponds to an average pH around the DNA/PEI polyplexes.

## 4.3 Results and Discussion

### *4.3.1 Quantitative Comparison of Intracellular Trafficking of Polyplexes in the Absence/Presence of Free PEI Chains in Clathrin-Mediated Pathway*

To address the effect of length and topology of free polycationic chains on nucleic acid delivery, we used a combination of two different PEI chains: one long PEI chain (*b*PEI-25K) to complex and encapsulate DNA and the other with a different length/topology as free chains. The choice of *b*PEI-25K as bound chains is due to its higher binding affinity to DNA, which provides better protection to the DNA payload in the presence of other long anionic chains, such as RNA and proteins, in the intracellular environment [41, 42]. Moreover, it is previous demonstrated that there is no interchange between the long *b*PEI-25K chains within the polyplexes and short branched or linear PEI chains free in the solution mixture, which allows us to focus on the impact of free chains only [24].

Figure 4.1b shows the effect of length and topology of free PEI chains on the gene transfection efficiency of DNA/*b*PEI-25K polyplexes in HepG2 cells, where $t = 0$ denotes a simultaneous addition of polyplexes (N:P = 3) and 7-portion free PEIs; and $t > 0$ represents that free PEI chains were applied hours after the administration of the polyplexes. As expected, the addition of free PEI chains can essentially enhance the transfection efficiency, largely independent of the chain length, topology (branched or linear), and time of addition ($t_{addtion}$). However, long free cationic chains ($\sim$25 K) are typically 2–20 times more efficient than their short counterparts, highlighting the length of free cationic chains plays a pivotal role in the gene transfection. Meanwhile, it is worth noting that for long free chains ($\sim$25 K), the chain topology has nearly no impact on the transfection efficiency; but for short free chains ($\sim$2 K), *lin*PEI-2.5K are $\sim$10-fold more effective than their branched counterpart, *b*PEI-2K. Note that similar trend is also revealed in other cells lines, such as 293T and HeLa cells.

In addition to the gene transfection efficacy, it was suggested that the cytotoxicity of free PEI are also strongly associated with its chain length and topology

[27, 28, 43, 44]. Thus, we evaluated the cytotoxicity and membrane leakage induced by different free PEIs in HepG2 cells, respectively, by using MTT and LDH assays. Figure 4.2 shows that (1) at $C_{PEI} \sim 3$ μg/mL, corresponding to N:P ~ 10 in the gene transfection, all of the free PEIs exhibited relatively low toxicity (cell viability well above 70 %), and only *b*PEI-25K chains mildly compromised the membrane integrity; and (2) when $C_{PEI} > 3$ μg/mL, long PEI chains (~25 K) are much more toxic to the cells, and in particular, branched long chains mediate significantly higher cytotoxicity than their linear counterparts, presumably due to their destabilization and disruption to the cellular membrane, as shown in Fig. 4.2b.

Since long free PEI chains are superior to short ones in delivering the DNA cargo, it was first investigated whether this is due to their promotion to the cell association and internalization of polyplexes, or alternatively, to the subsequent intracellular trafficking. Given that the cellular internalization of polyplexes almost reaches a maximum plateau and stops after ~4 h [24], those polyplexes in the extracellular space, if any, should remain there. If free PEI chains exclusively promote this cellular internalization process, the removal of polyplexes before the addition of free PEI chains should significantly reduce the transfection efficiency. However, Fig. 4.3 shows that when free PEIs, such as *b*PEI-25K chains, were

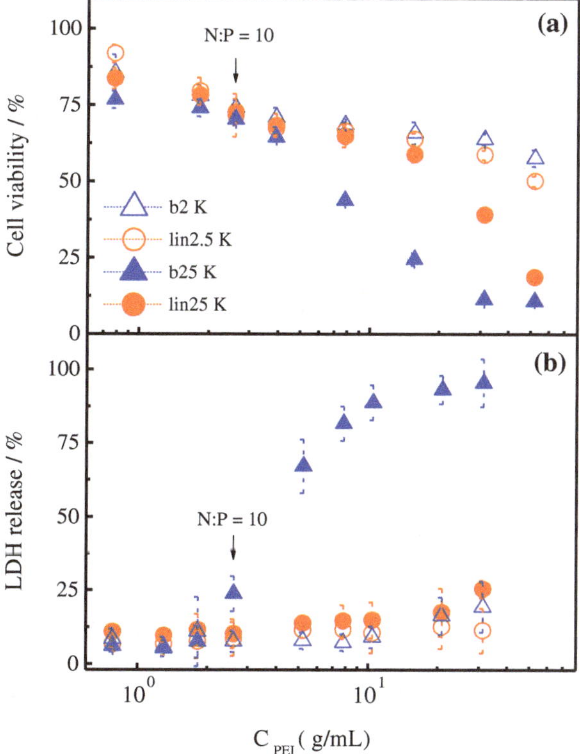


**Fig. 4.2** Concentration dependence of **a** HepG2 cell viability and **b** LDH release from HepG2 cells upon the treatment of free PEI chains with different length and topology, respectively, where $C_{PEI} = 2.7$ μg/mL corresponds to N:P = 10 in an effective gene transfection

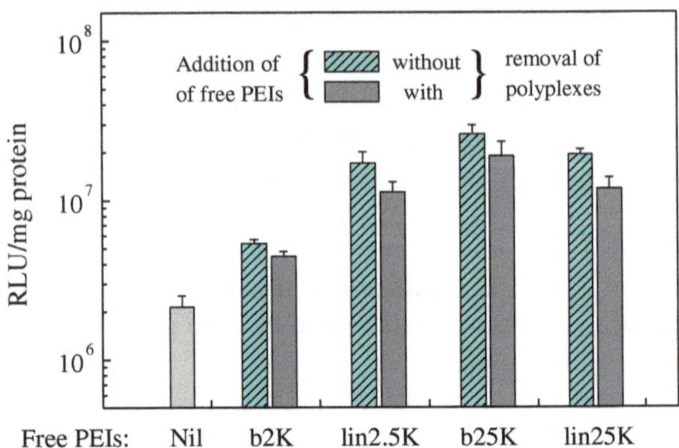


**Fig. 4.3** Effect of free PEI chains on the intracellular trafficking of DNA/*b*PEI-25K polyplexes (N:P = 3) in HepG2 cells, where 7 portions of different free PEIs were added, respectively, at 4 h post-administration of the polyplexes (N:P = 3), with or without the removal of cell culture medium. "Nil" means that no free PEIs were added

applied 4-h after polyplex administration, the gene transfection efficiency in polyplex-removal group was only attenuated by ~2 times compared to the non-removal group, implying that those free PEI chains promote the cellular internalization a little. Meanwhile, this reduced transfection efficiency is still ~20 times higher than that without the addition of free PEI chains (N:P = 3), clearly indicating that free PEI chains mainly play their role inside the cells.

Currently, it remains a challenge to elucidate how these free polycationic chains facilitate the intracellular trafficking of polyplexes since direct observation of their intracellular trafficking between different organelles is rather difficult. To visualize them, DNA was covalently labeled with Cy3 ($\lambda_{ex} = 550$ nm, $\lambda_{em} = 570$ nm, red), while lysosomes were labeled with a specific marker, Lamp-1 GFP ($\lambda_{ex} = 488$ nm, $\lambda_{em} = 520$ nm, green), via virus-mediated transduction. Herein, a confocal image-assisted three-dimensionally integrated quantification (CIQ) method was developed and employed to in situ monitor the subcellular localization of Cy3-DNA/*b*PEI-25K polyplexes upon the treatment of different free PEI chains. As shown in Fig. 4.4, at each indicated time, a series of Z-scan images were captured from each single cell, where the sum of Cy3-fluorescence intensity of yellow clusters (overlay of red and green) in all these Z-scan images ($I_{lyso}$) indexes *p*DNA content entrapped into lysosomes; and sum of Cy3-intensity of both red and yellow clusters ($I_{tot}$) indexes total pDNA internalized inside each cell. To optimize the number of Z-steps ($n_z$), we studied the effect of $n_z$ on the average Cy3-fluorescence intensity of DNA/PEI polyplexes ($I_{avg}$) in a fixed volume in HepG2 artificial cytosol. Figure 4.5a shows that $I_{avg}$ is nearly independent of $n_z$ when $n_z > 10$. Thus, n is kept to be 10 in the following experiments. Moreover, Fig. 4.5b shows that $I_{avg}$ linearly increases with the DNA concentration ($C_{DNA}$) when $n_z = 10$, so that the total uptake amount and fraction of Cy3-DNA entrapped into lysosomes ($F_{lyso}$) can be estimated from $m_{DNA} = C_{DNA} \times V_{cell}$ and $F_{lyso} = I_{lyso}/I_{tot} \times 100\ \%$, respectively, where $V_{cell}$ is the cell volume.

In terms of cellular uptake, Fig. 4.6 shows that long free PEI chains can slightly boost the uptake rate constant as well as the ultimate uptake amount, while short free chains have no such effect. At 6-h post-incubation, the final uptake amount in

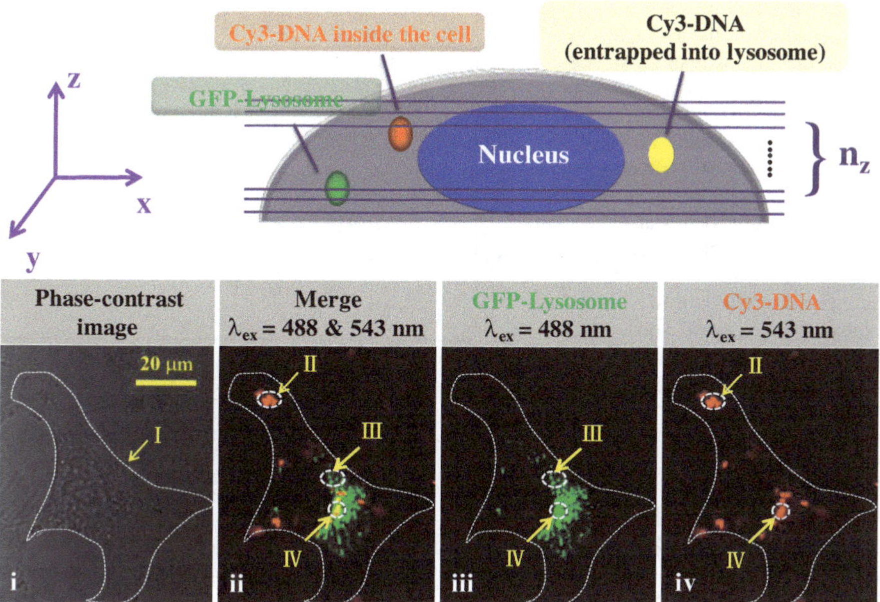


**Fig. 4.4** Schematic illustration of a confocal image-assisted three-dimensionally integrated quantification methodology to quantify the cellular internalization and fraction of DNA/PEI polyplexes entrapped into lysosomes ($F_{lyso}$), where DNA and lysosomes were labeled with Cy3 (*red*) and Lamp-1 GFP (*green*, indicated by *arrow III*), respectively. At each indicated time, 10 Z-scan images with a 1-μm step were captured from each cell. Sum of Cy3-intensity of yellow clusters (overlay of *red* and *green*, indicated by *arrow IV*) in ten images ($I_{lyso}$) indexes pDNA content entrapped into lysosomes; whereas sum of Cy3-fluorescence intensity of both *red* and *yellow* clusters ($I_{tot}$, indicated by *arrow II* and *IV*) indexes total pDNA inside each cell. It is demonstrated in Fig. 4.5b that average Cy3-fluorescence intensity per cell ($I_{avg}$) linearly increases with DNA concentration ($C_{DNA}$) so that amounts of Cy3-DNA inside cell and lysosomes are estimated from $m_{DNA} = C_{DNA} \times V_{cell}$ and $F_{lyso} = I_{lyso}/I_{tot} \times 100$ %, respectively, where $V_{cell}$ is the cell volume, and *arrow I* denotes the cellular membrane in each XY-plane

the presence of free *b*PEI-25K is ~1.2 × $10^5$ DNA copies/cell, only ~1.3-fold higher than that without free PEIs. On the other hand, the transgene expression at N:P = 10 is ~$10^2$-fold higher than that at N:P = 3, further indicating that these free cationic PEI chains mainly facilitate the intracellular journey of the DNA payload.

In the endolysosomal pathway, it was quantified that with the aid of long free *b*PEI-25K chains, the fraction of Cy3-DNA entrapped into lysosomes ($F_{lyso}$) slowly increases to ~20 % after 3 h but slightly decreases to ~15 % after 6 h. In contrast, without free PEI chains, the fraction of Cy3-DNA inside lysosomes keeps escalating and reaches ~40 % in the first 6 h (Fig. 4.7). These results are in line with the study of the intracellular pH variation around polyplexes, as shown in Fig. 4.8. Namely, without free PEI chains (N:P = 3), the intracellular pH around the polyplexes decreases from ~7.4 to ~5.9 in the first 6 h, whereas in the

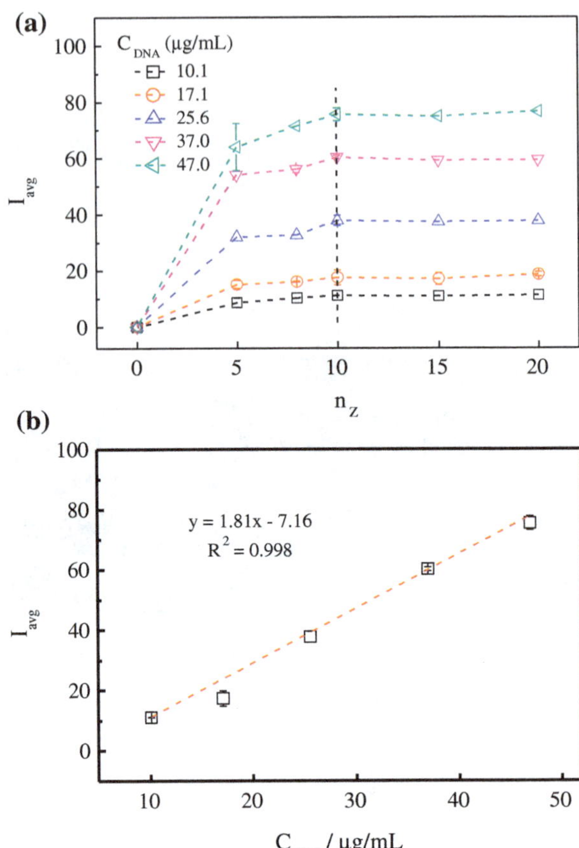


**Fig. 4.5** **a** Effect of number of Z-steps ($n_z$) on the average fluorescence intensity of Cy3-DNA/PEI ($I_{avg}$) in a fixed volume of HepG2 artificial cytosol. For preparation of artificial cytosol, $10^6$ HepG2 cells per mL of PBS was sonicated for ten rounds at 4 °C by using an ultrasonic cell crusher machine and then centrifuged at 6000x g to remove the cell nuclei and debris. The confocal scanning range in the Z-direction is fixed to be 10 μm, and the area in XY-plane is $9.7 \times 10^4$ μm$^2$. **b** DNA concentration ($C_{DNA}$) dependence of $I_{avg}$ in a fixed volume of HepG2 artificial cytosol, where $n_z$ is kept to be 10 and the interval of each Z-step is 1 μm

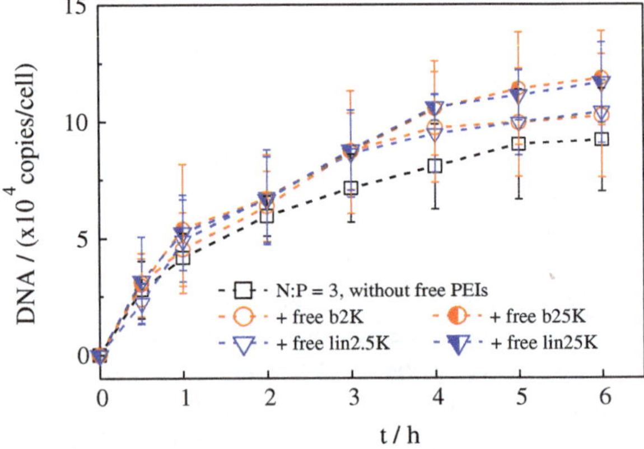


**Fig. 4.6** Effect of length and topology of free PEI chains on the cellular uptake of Cy3-DNA/*b*PEI-25K polyplexes (N:P = 3) monitored by using a CIQ method, where the total and final N:P is kept to be 10, and at least 10 cells were analyzed in each case

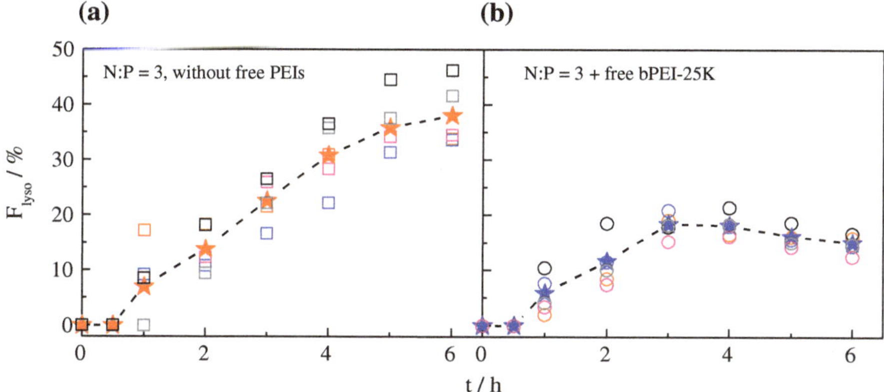


**Fig. 4.7** Effect of 7 portions of long free *b*PEI-25K chains on the fraction of Cy3-DNA/*b*PEI-25K polyplexes (N:P = 3) entrapped in lysosomes per HepG2 cell ($F_{lyso}$) monitored by using a CIQ method, where 5 cells were analyzed in each experimental condition

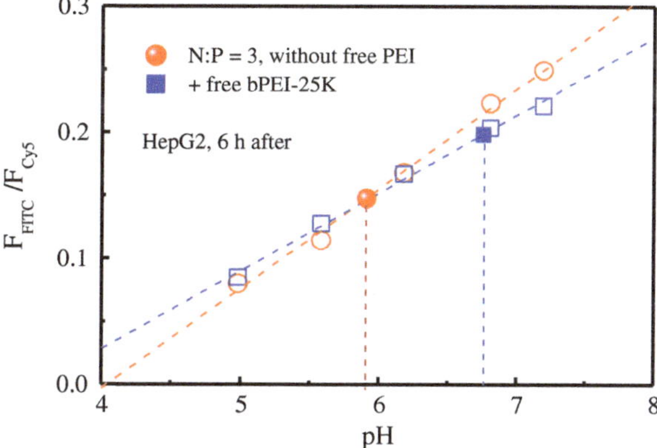


**Fig. 4.8** pH calibration curves (*hollow symbols*), where each symbol represents one study of DNA/*b*PEI-25K polyplexes (N:P = 3) without/with 7 portions of long free *b*PEI-25K chains, and *filled symbols* denote the intracellular pH near the polyplexes in each case, which were respectively obtained from the normalized relative fluorescence intensity via each calibration curve

presence of long free *b*PEI-25K chains (N:P = 10), the pH value only slightly decreases to ~6.8. A combination of these results quantitatively indicates that long free PEI chains are able to prevent the entrapment of polyplexes into the acidic and degradative lysosomes, presumably by (1) facilitate the release of polyplexes from the early endosomes or even from the small virginal endocytic vesicles; and more importantly, (2) hamper or delay the development of the later endolysosomes.

Further, we compared the efficacy of different free PEI chains in preventing the entrapment of polyplexes into lysosomes. Notably, in the presence of short branched PEI chains, the lysosomal distribution of polyplexes followed the similar trend to that without free chains, whereas with the aid of short linear free PEI chains, the amount of polyplexes entrapped into the lysosomes was significantly lower (Fig. 4.9a). On the other hand, long free PEIs are shown to greatly reduce

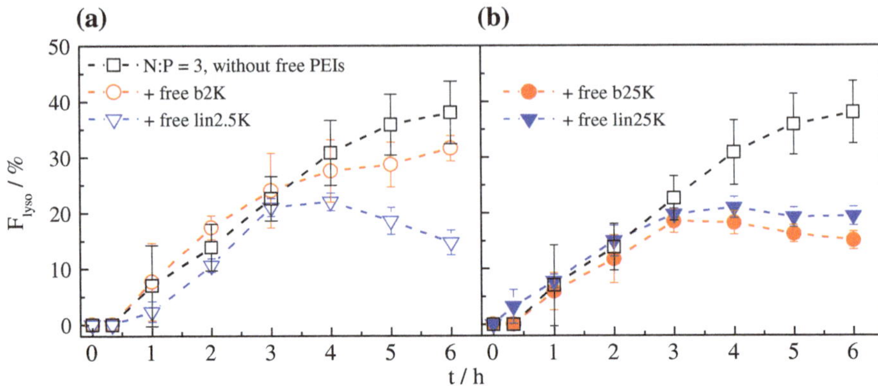


**Fig. 4.9** Effect of topology (branched or linear) of **a** short (~2 K) and **b** long (~25 K) free PEI chains on fraction of Cy3-DNA/*b*PEI-25K polyplexes (N:P = 3) entrapped into lysosomes per HepG2 cell ($F_{lyso}$), where a CIQ method was used and at least 5 cells were analyzed in each experimental condition

the lysosomal entrapment of polyplexes from ~40 to ~15 % at 6 h post-transfection, irrespective of the chain topology (Fig. 4.9b).

Collectively, the results listed above are well consistent with and partially validate our hypothesis on why long free PEI chains are more potent in facilitating the intracellular trafficking of the DNA payload [24]. Study on internalization and intracellular trafficking of long FITC-PEIs using real-time CLSM shows that within the first 30 min, long free *b*PEI-25K chains mainly located and sometimes formed clusters in the cell membrane. After ~2 h, majority of free PEI chains accumulated in the organelles in the cytoplasm, and particularly, some of them co-localized with the late endosomes/lysosomes (Fig. 4.10).

It is postulated that it is those long free cationic PEI chains (~15–20 nm) embedded inside the anionic cell membrane (~5–6 nm) via electrostatic interaction that (1) destabilize/disrupt the membrane of endosomes or even the virginal endocytic vesicles (the "proton sponge" effect might help here) and trigger the escape of the polyplexes residing inside; and (2) prevent the fusion between the endosomes and lysosomes so that further development of the later endolysosomes is delayed/inhibited. It is suggested that the stick-out cationic chain end(s) might effectively interact with or shield those "signaling" anionic proteins, which are attached to the inner surface of the cell membrane (i.e., the outer surface of the endosomes), for inter-vesicle fusion. Consequently, many of the virginal endocytic vesicles with polyplesxes inside do not fuse with the later endolysosomes compartment. Using this hypothesis, we can explain many differences observed in gene transfection studies by us and others, such as (1) why short branched PEI chains (e.g., *b*PEI-0.8K ~4 nm and *b*PEI-2K ~6 nm) suffer from low transfection efficacy because they might be too short to shield the membrane proteins served as a signal for inter-vesicular fusion; (2) why short linear PEI (e.g., *lin*PEI-2.5K) is more potent than its branched counterpart in preventing the entrapment of

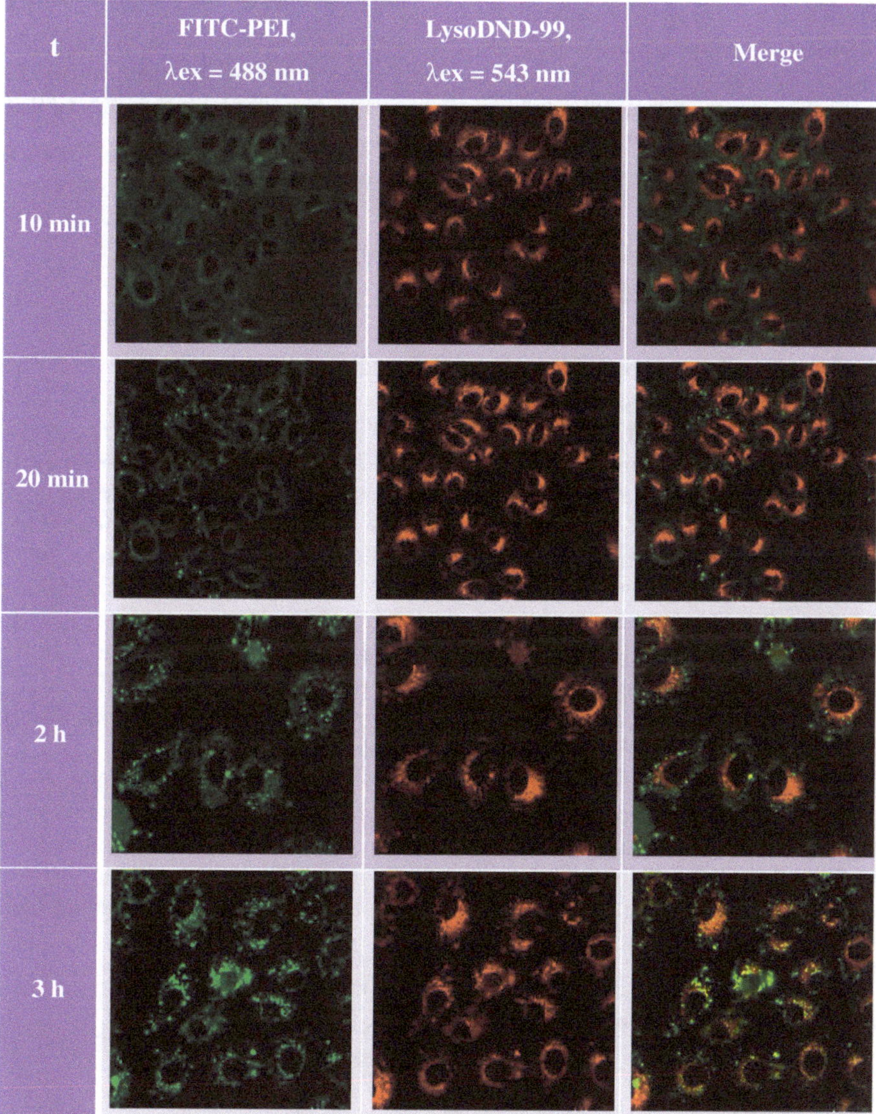


**Fig. 4.10** Translocation of FITC-labeled *b*PEI-25K chains into HepG2 cells and subsequent intracellular trafficking, where time (t) is counted after addition of FITC-PEI, CPEI = 2.7 μg/mL in serum-free DMEM, and FITC-PEI and lysoDND 99 (primarily staining the late endosomes/lysosomes) were visualized by excitation with 488- and 543-nm laser, respectively

polyplexes into the lysosomes? (*lin*PEI-2.5K chain is ~18 nm if stretched, much longer than *b*PEI-2K chain); (3) why *lin*PEI-25K and *b*PEI-25K are similarly efficient in preventing the lysosomal entrapment even though linear chains have a much longer contour length? (*lin*PEI-25K chains are coiled in solution so that they have a similar size as the branched ones); and (4) why attaching cholesterol [45] or a hydrophobic aliphatic lipid with proper length (e.g., palmitic acid, $C_{15}H_{31}CO$–) [46] to the less effective *b*PEI-2K can significantly improve its transfection efficacy? (The short aliphatic lipid chain can insert into the membrane so that *b*PEI-2K can stick out to shield the signal proteins).

## 4.3.2 Effect of Endocytosis Pathway on the Intracellular Fate of Polyplexes

In addition to the well-documented clathrin-mediated (CME) pathway, we further investigated the effect of a clathrin-independent route, caveolae-mediated pathway, on the intracellular fate of polyplexes, and particularly, explored the possibly role of free PEI chains in these two routes. To address this issue, we choose to use HeLa cell line, which can take up the polyplexes via both the CME and caveolae-mediated endocytosis [34, 35]. Figure 4.11 shows that the efficacy of different free PEI chains in promoting the gene transfection in HeLa cells follows *b*PEI-25K > *lin*PEI-2.5K > *b*PEI-2K, consistent with our previous results in 293T and HepG2 cell lines [24].

Further, we determined the optimal concentration of each specific inhibitor targeting the CME or caveolae-mediated pathway to maximize its efficacy and minimize its cytotoxicity (Table 4.1). The concentration of chlorpromazine is determined to be ~21 μM, at which the uptake of Alexa-488-transferrin ($T_f$), a ligand internalized exclusively via CME pathway, was inhibited by over 90 % assayed using flow cytometry (data not shown). Likewise, the concentration of filipin III and genistein are ~3.8 and 100 μM, respectively, at which the uptake of Alexa-488-toxin, a ligand internalized exclusively via caveolae pathway, was blocked by over 90 %. Figure 4.12 shows that each inhibitor stimulates minimal

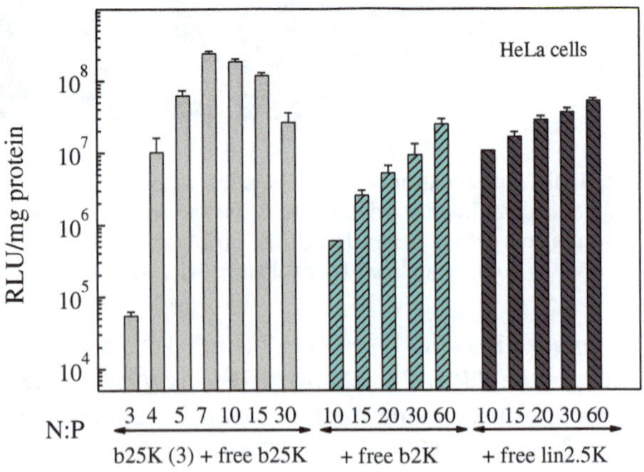


**Fig. 4.11** Evaluation on the efficacy of free PEI chains with different length/topology in enhancing the gene transfection efficiency of DNA/*b*PEI-25K polyplexes (N:P = 3, denoted as b25K (3)) in HeLa cells, where "N:P" denotes the total and final N:P ratio of the polyplexes plus free PEI chains

**Table 4.1** Inhibitors Used and Corresponding Targeted Endocytosis Pathways

| Inhibitors | Targeted pathway | Inhibition mechanism | Concentration (μM) |
|---|---|---|---|
| Chlorpromazine | Clathrin-mediated | Dissociate clathrin lattice | 21 |
| Filipin III | Caveolae-mediated | Sterol-binding agent, removes cholesterol from plasma membrane | 3.8 |
| Genistein | Caveolae-mediated | Tyrosine kinase inhibitor | 100 |

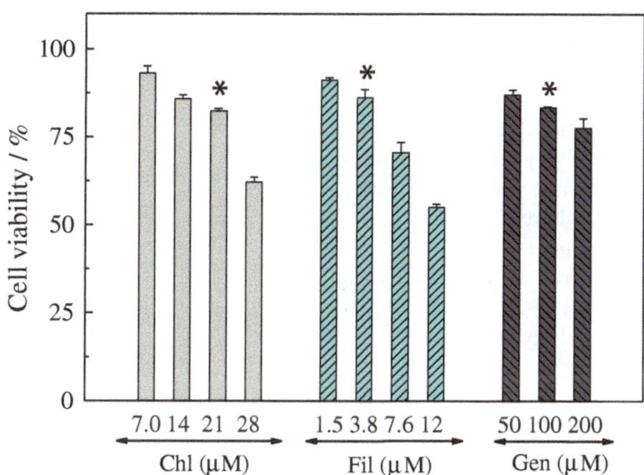


**Fig. 4.12** Concentration dependence of different specific endocytosis inhibitors on the HeLa cell viability, where "*" denotes that cell viability is >80 % at the concentration used for the following gene transfection tests. "Chl", "Fil" and "Gen" refer to chlorpromazine, filipin III and genistein, respectively

toxicity at its corresponding applied concentration, with the cell viability well above 80 %.

It is previously demonstrated that blocking of either the CME or caveolae pathway in HeLa cells would lead to ~50 % decrease in the cellular uptake of DNA/*b*PEI-25K polyplexes, and these two endocytosis routes might be sometimes interchangeable [35]. Thus, in the following studies, we mainly investigated the effect of blocking each pathway on the ultimate gene transfection efficiency of polyplexes (N:P = 3), in the presence of limited or sufficient amount of free PEI chains. Figure 4.13a shows that at N:P = 4, with limited amount of free *b*PEI-25K chains, the transfection efficiency was nearly completely lost when caveolae-mediated pathway was blocked. This is because in CME pathway such a small amount of free PEI chains cannot prevent the entrapment of polyplexes into the late endosomes/lysosomes so that most of the polyplexes are destined to lysosomes and suffer from degradation. Alternatively, the transfection efficiency was enhanced by ~3 times when CME pathway was inhibited, suggesting that at lower N:P ratios (N:P $\leq$ 4) the caveolae-dependent route is more favorable for an effective gene transfection due to the relatively low possibility of caveolar vesicles to fuse with lysosomes [20, 47]. On the other hand, at N:P = 7, with sufficient amount of free PEI chains, the inhibition of either the CME or caveolae pathway only decreased the gene transfection efficiency by ~2 times, implying that long

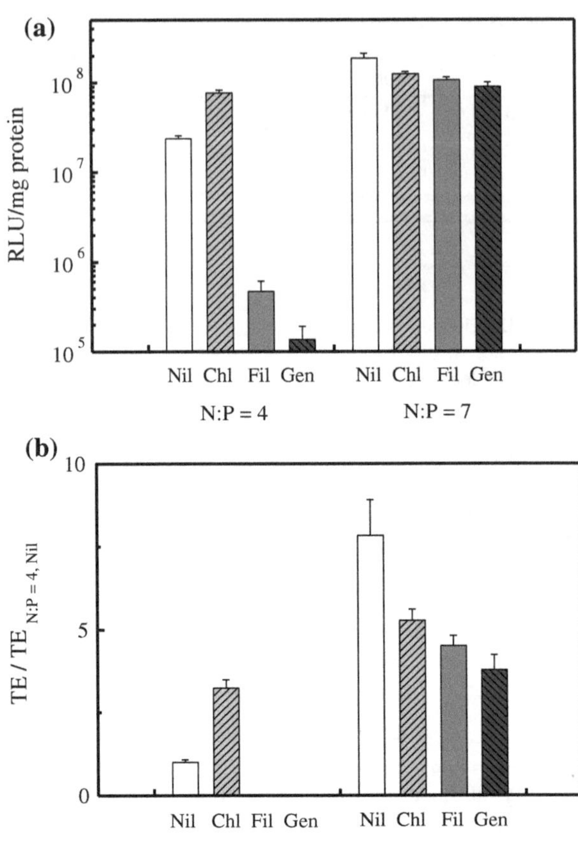


**Fig. 4.13** **a** Effect of treatment of different specific endocytosis inhibitors on the transfection efficiency in HeLa cells, where "N:P = 4" and "N:P = 7" denotes DNA/*b*PEI-25K polyplexes (N:P = 3) plus 1 and 4 portions of free *b*PEI-25K chains, respectively. "Nil" means that no inhibitor was added, while "Chl", "Fil" and "Gen" represent chlorpromazine (21 μM), filipin III (3.8 μM) and genistein (100 μM), respectively. **b** Normalized gene transfection efficiency (TE), where $TE_{N:P = 4, Nil}$ represents the TE when the HeLa cells were treated with DNA/*b*PEI-25K of N:P = 4 in the absence of any inhibitors

cationic free PEI chains are able to (1) prevent the entrapment of polyplexes into lysosomes via the perturbation of the endosome-to-lysosome fusion in CME pathway; and (2) facilitate the release of polyplexes from the virginal endocytic or caveolar vesicles.

## 4.4 Conclusion

A stringent of quantitative methods, including real-time confocal three-dimensionally integrated quantification, flow cytometry and specific endocytosis inhibitor treatment tests, were designed and employed to address how free PEI chains with a proper length/topology promote the endocytosis and the subsequent trafficking, particularly in the clathrin-mediated (CME) pathway. Cellular uptake kinetics reveals that free cationic PEI chains, especially those long ones (~25 K), are able to quickly penetrate different membranes of the cell, and meanwhile, boost the uptake rate constant of the polyplexes, presumably due to their disruptive nature to the anionic cell membrane. Nevertheless, the major role played by free PEI chains is in the intracellular space. Particularly, in the CME pathway, the efficacy of free cationic PEI chains is tightly correlated to their ability in preventing the entrapment of polyplexes into the late endosomes/lysosomes. Namely, without free PEI chains or with short branched ones (~2 K), many of the polyplexes (>40 % in HepG2 cells at 6-h post-administration) are destined to the acidic lysosomes and suffer from degradation. On the other hand, the addition of either the short linear (*lin*PEI-2.5K) or long free chains (~25 K) effectively facilitates the intracellular trafficking, with less than 15 % of the internalized polyplexes trapped inside lysosomes.

In addition to the CME pathway, the efficacy of long cationic free PEI chains in promoting the intracellular trafficking is also revealed in the caveolae-mediated pathway in HeLa cells. At N:P = 4, with limited amount of free PEIs, the gene transfection was almost completely abolished when the caveolae-mediated pathway was blocked, whereas the transfection efficiency was enhanced by ~3 times when the CME pathway was inhibited. This indicates that at lower N:P ratios (N:P $\leq$ 4) most of the polyplexes cannot survive in the CME pathway, and only the caveolae-dependent route leads to an efficient gene transfection because there is a relatively less chance for caveolar vesicles to fuse with the late endosomes/ lysosomes. On the other hand, at N:P = 7, with sufficient amount of long free PEI chains, the inhibition of either the CME or caveolae pathway only decreased the gene transfection efficiency by ~2-fold, and this decrease is possibly due to the reduced cellular uptake.

Collectively, the results presented in this chapter are well consistent with and can partially validate our hypothesis on why long free cationic PEI chains are more potent in facilitating the intracellular trafficking of the DNA cargo [24]. It is postulated that free cationic PEI chains with a proper length (~15–20 nm) embedded inside the anionic cell membrane are able to (1) compromise/disrupt the membrane of the virginal endocytic vesicles, endosomes or caveosomes (the "proton sponge" effect might be helpful) and trigger the escape of the polyplexes entrapped inside; and (2) block the endocytic-vesicle-to-endolysosome fusion via interacting with or shielding the anionic signaling proteins for inter-vesicular fusion, so that further development of the later endolysosomes is prohibited or delayed.

## 4.5 Future Research and Perspective on Development of Non-Viral Polymeric Vectors

Over the past few decades, polycationic chains with different sizes and topologies, sometimes exotic, have been designed and synthesized to deliver genes into a variety of cells, organs and tissues. However, in most of the cases, their transfection efficacy remains disappointing, orders of magnitude lower than their viral counterparts. It is my opinion that this is, at least partially, due to the lack of some fundamental understanding on how DNA is delivered into the cell nucleus; namely, a detailed pathway for the intracellular trafficking of the polyplexes. In the last 15 years, the astonishing advancement of molecular cell biology and its related commercially available analytic tools and kits have now enabled us to address this complicated question. The future research on development of non-viral polymeric vectors might include the following important issues.

(1) Much attention has already been paid to endocytosis in the past, especially the DNA complexation and encapsulation in the extracellular space. More and more evidence has been accumulated to show that the complexation between the anionic DNA and cationic polymer chains is simply due to the charge neutrality and is a counterion-related entropy driven process [48]. Therefore, we should shift our attention away from it. Note that ingesting part of the plasma membrane (endocytosis) is a constant cell activity. It normally takes 1–2 h for a typical cell to replace its entire plasma membrane via the endo-and-exo-cytosis circle. Thus, the endocytosis of the polyplexes with appropriate cell-targeting ligands should not be a great hurdle as long as we are able to bring them sufficiently close to the cell membrane. This is why a slightly positively charged periphery of the engineered polyplexes is important and the attachment of proper ligands to target receptors on cell membrane is helpful.
(2) Much attention has also been paid to the escape of polyplexes from the endolysosomes by using polycationic chains that have a buffer capacity or are pH-sensitive. Recent experimental results have revealed that it is those polycationic chains free in the solution mixture (N:P $\geq$ 6) that actually promote the gene transfection [23, 24, 26–28], presumably via (1) destabilize/disrupt the membrane of the virginal endocytic vesicles, endosomes or caveosomes (the "proton sponge" effect might play some roles here) and trigger the escape of the polyplexes entrapped inside; and (2) block the virginal-vesicle-to-endolysosome pathway via shielding or malfunctioning the anionic signaling proteins attached at the inner surface of the cell membrane (i.e., the outer surface of the endocytic vesicles). Thus, in addition to escaping from endolysosomes by virtue of the "proton sponge" effect, we should also concentrate on how to block the inter-vesicular fusion between the ingested polyplexes-containing virginal vesicles and early endosomes so that they will not be developed into the later endolysosomes. In this way, the escape of trapped polyplexes from lysosomes will not be an issue. The important related two issues here will be the detailed molecular mechanisms of (a) how free polycationic chains prevent the inter-vesicular fusion; and (b) how the polyplexes escape from the small ingested virginal vesicles into the cytosol. To elucidate these two issues, we plan to improve the spatio-temporal observation resolution by using spinning disk CLSM, which allows acquisition of images at very high frame rates with minimum illumination to the samples [49]. Such high-frame-rate observation will reveal much more detailed information on the

transport and fusion of the polyplex-containing vesicles among different organelles, from which we can identify the key events for an efficient intracellular trafficking. Alternatively, RNA interference (RNAi) will be employed to suppress the expression of several possible proteins responsible for intervesicular fusion. Rab5, one of such endocytic proteins, is known as an important factor required for the homotypic fusion between early endosomes and the heterotypic fusion between early endosomes and the virginal plasma-membrane-derived vesicles [50, 51]. Thus, we plan to knockdown one or two Rab5 isoforms to attenuate/prevent the fusion between polyplex-containing vesicles and late endosomes/lysosomes, and at the same time, ensure this knockdown does not greatly impair the regular cell internalization.

(3) More attention should be given to the transport of polyplexes through the cytosol. Note that the cytosol is a fairly concentrated protein solution ($\sim$30 % by weight) with a high viscosity. It is hard to imagine that the polyplexes are able to passively move towards the cell nuclear membrane by thermal diffusion. Some past experiments showed that polyplexes within endocytic vesicles actively travel along the microtubules and are powered by ATP [52, 53]. More studies and convincing evidence are required to confirm such a pathway even though the migration of the polyplexes towards the cell nucleus might not be a rate-determining step but is certainly important.

(4) More efforts should be directed to a better understanding of how the polyplexes or large DNA chains actually pass through the nuclear membrane, especially when the cells are not in their mitosis state. Few subsequent associated questions are as follows: (a) are DNA chains released from the polyplexes before or after passing through the nuclear membrane?; (b) how are the polyplexes or even the released DNA chains able to pass through the nuclear pores much smaller than them?; and (c) how can we artificially induce the temporal formation of large pores on the nuclear membrane or the temporal dismantlement of the nuclear membrane to allow the polyplexes or released DNA chains into the nucleus.

(5) Chemists should learn from molecular cell biologists to understand and find some subtle differences between micro-environments in the cytosol and the cell nucleus so that one can use them to design and prepare a new generation of efficient non-viral polymeric vectors for the gene transfection; namely, these novel vectors can release their captured DNA chains in a more controllable fashion in the intracellular space. The problem is extremely complicated and multidimensional and so do its researchers. Polymer researchers, who are interested in the development of useful, efficient non-viral vectors, have no choice but sit down to learn sufficient molecular cell biology and pharmacology because our future is multidisciplinary.

## References


1. Mintzer, M. A., & Simanek, E. E. (2009). Nonviral vectors for gene delivery. *Chemical Reviews, 109*, 259–302.
2. Canine, B. F., & Hatefi, A. (2010). Development of recombinant cationic polymers for gene therapy research. Also change "Journal of Controlled Release" to "Advance Drug Delivery Review". *Journal of Controlled Release, 62*, 1524–1529.
3. Won, Y., Lim, K. S., & Kim, Y. (2011). Intracellular organelle-targeted non-viral gene delivery systems. *Journal of Controlled Release, 152*, 99–109.
4. Pack, D. W., Hoffman, A. S., Pun, S., & Stayton, P. S. (2005). Design and development of polymers for gene delivery. *Nature Review Drug Discovery, 4*, 581–593.
5. Li, S. D., & Huang, L. (2007). Non-viral is superior to viral gene delivery. *Journal of Controlled Release, 123*, 181–183.
6. Boussif, O., Lezoualch, F., Zanta, M. A., Mergny, M. D., Scherman, D., Demeneix, B., et al. (1995). A versatile vector for gene and oligonucleotide transfer into cells in culture and in vivo: polyethylenimine. *Proceedings of the National Academy of Sciences of the United States of America, 92*, 7297–7301.
7. Lungwitz, U., Breunig, M., Blunk, T., & Gopferich, A. (2005). Polyethylenimine-based non-viral gene delivery systems. *European Journal of Pharmaceutics and Biopharmaceutics, 60*, 247–266.
8. Neu, M., Fischer, D., & Kissel, T. (2005). Recent advances in rational gene transfer vector design based on poly(ethylene imine) and its derivatives. *Journal of Gene Medicine, 7*, 992–1009.
9. Ogris, M., Brunner, S., Schuller, S., Kircheis, R., & Wagner, E. (1999). PEGylated DNA/transferrin-PEI complexes: reduced interaction with blood components, extended circulation in blood and potential for systemic gene delivery. *Gene Therapy, 6*, 595–605.
10. Cheng, H., Zhu, J. L., Zeng, X., Jing, Y., Zhang, X. Z., & Zhuo, R. X. (2009). Targeted gene delivery mediated by folate-polyethylenimine-block-poly(ethylene glycol) with receptor selectivity. *Bioconjugate Chemistry, 20*, 481–487.
11. Zanta, M. A., Boussif, O., Adib, A., & Behr, J. P. (1997). In vitro gene delivery to hepatocytes with galactosylated polyethylenimine. *Bioconjugate Chemistry, 8*, 839–844.
12. Diebold, S. S., Kursa, P., Wagner, E., Cotten, M., & Zenke, M. (1999). Mannose polyethylenimine conjugates for targeted DNA delivery into dendritic cells. *Journal of Biological Chemistry, 274*, 19087–19094.
13. Gosselin, M. A., Guo, W. J., & Lee, R. J. (2001). Efficient gene transfer using reversibly cross-linked low molecular weight polyethylenimine. *Bioconjugate Chemistry, 12*, 989–994.
14. Thomas, M., Ge, Q., Lu, J. J., Chen, J. Z., & Klibanov, A. M. (2005). Cross-linked small polyethylenimines: while still nontoxic, deliver DNA efficiently to mammalian cells in vitro and in vivo. *Pharmaceutical Research, 22*, 373–380.
15. Breunig, M., Lungwitz, U., Liebl, R., & Goepferich, A. (2007). Breaking up the correlation between efficacy and toxicity for nonviral gene delivery. *Proceedings of the National Academy of Sciences of the United States of America, 104*, 14454–14459.
16. Deng, R., Yue, Y., Jin, F., Chen, Y. C., Kung, H. F., Lin, M. C. M., et al. (2009). Revisit the complexation of PEI and DNA - how to make low cytotoxic and highly efficient PEI gene transfection non-viral vectors with a controllable chain length and structure? *Journal of Controlled Release, 140*, 40–46.
17. Peng, Q., Zhong, Z. L., & Zhuo, R. X. (2008). Disulfide cross-linked polyethylenimines (PEI) prepared via thiolation of low molecular weight PEI as highly efficient gene vectors. *Bioconjugate Chemistry, 19*, 499–506.
18. Kang, H. C., Kang, H. J., & Bae, Y. H. (2011). A reducible polycationic gene vector derived from thiolated low molecular weight branched polyethyleneimine linked by 2-iminothiolane. *Biomaterials, 32*, 1193–1203.

19. Bae, Y. M., Choi, H., Lee, S., Kang, S. H., Kim, Y. T., Nam, K., et al. (2007). Dexamethasone-conjugated low molecular weight polyethylenimine as a nucleus-targeting lipopolymer gene carrier. *Bioconjugate Chemistry, 18*, 2029–2036.
20. Won, Y. Y., Sharma, R., & Konieczny, S. F. (2009). Missing pieces in understanding the intracellular trafficking of polycation/DNA complexes. *Journal of Controlled Release, 139*, 88–93.
21. Medina-Kauwe, L. K., Xie, J., & Hamm-Alvarez, S. (2005). Intracellular trafficking of nonviral vectors. *Gene Therapy, 12*, 1734–1751.
22. Lechardeur, D., Verkman, A. S., & Lukacs, G. L. (2005). Intracellular routing of plasmid DNA during non-viral gene transfer. *Advanced Drug Delivery Review, 57*, 755–767.
23. Yue, Y., Jin, F., Deng, R., Cai, J., Chen, Y., Lin, M. C. M., et al. (2011). Revisit complexation between DNA and polyethylenimine – Effect of uncomplexed chains free in the solution mixture on gene transfection. *Journal of Controlled Release, 155*, 67–76.
24. Yue, Y., Jin, F., Deng, R., Cai, J. G., Dai, Z. J., Lin, M. C. M., et al. (2011). Revisit complexation between DNA and polyethylenimine – Effect of length of free polycationic chains on gene transfection. *Journal of Controlled Release, 152*, 143–151.
25. Clamme, J. P., Azoulay, J., & Mely, Y. (2003). Monitoring of the formation and dissociation of polyethylenimine/DNA complexes by two photon fluorescence correlation spectroscopy. *Biophysical Journal, 84*, 1960–1968.
26. Clamme, J. P., Krishnamoorthy, G., & Mely, Y. (2003). Intracellular dynamics of the gene delivery vehicle polyethylenimine during transfection: investigation by two photon fluorescence correlation spectroscopy. *Biochimica et Biophysica Acta, Biomembranes, 1617*, 52–61.
27. Boeckle, S., von Gersdorff, K., van der Piepen, S., Culmsee, C., Wagner, E., & Ogris, M. (2004). Purification of polyethylenimine polyplexes highlights the role of free polycations in gene transfer. *Journal of Gene Medicine, 6*, 1102–1111.
28. Fahrmeir, J., Gunther, M., Tietze, N., Wagner, E., & Ogris, M. (2007). Electrophoretic purification of tumor-targeted polyethylenimine-based polyplexes reduces toxic side effects in vivo. *Journal of Controlled Release, 122*, 236–245.
29. Saul, J. M., Wang, C. H. K., Ng, C. P., & Pun, S. H. (2008). Multilayer nanocomplexes of polymer and DNA exhibit enhanced gene delivery. *Advanced Materials, 20*, 19–25.
30. Forrest, M. L., Meister, G. E., Koerber, J. T., & Pack, D. W. (2004). Partial acetylation of polyethylenimine enhances in vitro gene delivery. *Pharmaceutical Research, 21*, 365–371.
31. Funhoff, A. M., van Nostrum, C. F., Koning, G. A., Schuurmans-Nieuwenbroek, N. M. E., Crommelin, D. J. A., & Hennink, W. E. (2004). Endosomal escape of polymeric gene delivery complexes is not always enhanced by polymers buffering at low pH. *Biomacromolecules, 5*, 32–39.
32. Dubruel, P., Christiaens, B., Vanloo, B., Bracke, K., Rosseneu, M., Vandekerckhove, J., et al. (2003). Physicochemical and biological evaluation of cationic polymethacrylates as vectors for gene delivery. *European Journal of Pharmaceutical Sciences, 18*, 211–220.
33. Dubruel, P., Christiaens, B., Rosseneu, M., Vandekerckhove, J., Grooten, J., Goossens, V., et al. (2004). Buffering properties of cationic polymethacrylates are not the only key to successful gene delivery. *Biomacromolecules, 5*, 379–388.
34. Gabrielson, N. P., & Pack, D. W. (2009). Efficient polyethylenimine-mediated gene delivery proceeds via a caveolar pathway in HeLa cells. *Journal of Controlled Release, 136*, 54–61.
35. Rejman, J., Bragonzi, A., & Conese, M. (2005). Role of clathrin- and caveolae-mediated endocytosis in gene transfer mediated by lipo- and polyplexes. *Molecular Therapy, 12*, 468–474.
36. van der Aa, M. A. E. M., Huth, U. S., Hafele, S. Y., Schubert, R., Oosting, R. S., Mastrobattista, E., et al. (2007). Cellular uptake of cationic polymer-DNA complexes via caveolae plays a pivotal role in gene transfection in COS-7 cells. *Pharmaceutical Research, 24*, 1590–1598.

37. Akita, H., Ito, R., Khalil, I. A., Futaki, S., & Harashima, H. (2004). Quantitative three-dimensional analysis of the intracellular trafficking of plasmid DNA transfected by a nonviral gene delivery system using confocal laser scanning microscopy. *Molecular Therapy, 9*, 443–451.
38. Hama, S., Akita, H., Ito, R., Mizuguchi, H., Hayakawa, T., & Harashima, H. (2006). Quantitative comparison of intracellular trafficking and nuclear transcription between adenoviral and lipoplex systems. *Molecular Therapy, 13*, 786–794.
39. Varga, C. M., Tedford, N. C., Thomas, M., Klibanov, A. M., Griffith, L. G., & Lauffenburger, D. A. (2005). Quantitative comparison of polyethylenimine formulations and adenoviral vectors in terms of intracellular gene delivery processes. *Gene Therapy, 12*, 1023–1032.
40. Akinc, A. (2002). Measuring the pH environment of DNA delivered using nonviral vectors: implications for lysosomal trafficking. *Biotechnology and Bioengineering, 78*, 503–508.
41. Itaka, K., Harada, A., Yamasaki, Y., Nakamura, K., Kawaguchi, H., & Kataoka, K. (2004). In situ single cell observation by fluorescence resonance energy transfer reveals fast intra-cytoplasmic delivery and easy release of plasmid DNA complexed with linear polyethylenimine. *Journal of Gene Medicine, 6*, 76–84.
42. Bertschinger, M., Backliwal, G., Schertenleib, A., Jordan, M., Hacker, D. L., & Wurm, F. M. (2006). Disassembly of polyethylenimine-DNA particles in vitro: implications for polyethylenimine-mediated DNA delivery. *Journal of Controlled Release, 116*, 96–104.
43. Moghimi, S. M., Symonds, P., Murray, J. C., Hunter, A. C., Debska, G., & Szewczyk, A. (2005). A two-stage poly(ethylenimine)-mediated cytotoxicity: implications for gene transfer/therapy. *Molecular Therapy, 11*, 990–995.
44. Parhamifar, L., Larsen, A. K., Hunter, A. C., Andresen, T. L., & Moghimi, S. M. (2010). Exploring polyethylenimine mediated DNA transfection and the proton sponge hypothesis. *Soft Matter, 6*, 4001–4009.
45. Wang, D. A., Narang, A. S., Kotb, M., Gaber, A. O., Miller, D. D., Kim, S. W., et al. (2002). Novel branched poly(ethylenimine)-cholesterol water-soluble lipopolymers for gene delivery. *Biomacromolecules, 3*, 1197–1207.
46. Neamnark, A., Suwantong, O., Bahadur, K. C. R., Hsu, C. Y. M., Supaphol, P., & Uludag, H. (2009). Aliphatic lipid substitution on 2 kDa polyethylenimine improves plasmid delivery and transgene expression. *Molecular Pharmaceutics, 6*, 1798–1815.
47. Pelkmans, L., Burli, T., Zerial, M., & Helenius, A. (2004). Caveolin-stabilized membrane domains as multifunctional transport and sorting devices in endocytic membrane traffic. *Cell, 118*, 767–780.
48. Bloomfield, V. A. (1997). DNA condensation by multivalent cations. *Biopolymers, 44*, 269–282.
49. Nakano, A. (2002). Spinning-disk confocal microscopy – a cutting-edge tool for imaging of membrane traffic. *Cell Structure and Function, 27*, 349–355.
50. Bucci, C., Parton, R. G., Mather, I. H., Stunnenberg, H., Simons, K., Hoflack, B., et al. (1992). The small GTPase rab5 functions as a regulatory factor in the early endocytic pathway. *Cell, 70*, 715–728.
51. Huang, F., Khvorova, A., Marshall, W., & Sorkin, A. (2004). Analysis of clathrin-mediated endocytosis of epidermal growth factor receptor by RNA interference. *Journal of Biological Chemistry, 16*, 16657–16661.
52. Suh, J., Wirtz, D., & Hanes, J. (2003). Efficient active transport of gene nanocarriers to the cell nucleus. *Proceedings of the National Academy of Sciences of the United States of America, 100*, 3878–3882.
53. Bausinger, R., von Gersdorff, K., Braeckmans, K., Ogris, M., Wagner, E., Brauchle, C., et al. (2006). The transport of nanosized gene carriers unraveled by live-cell imaging. *Angewandte Chemie International Edition, 45*, 1568–1572.

Zeitfracht Medien GmbH
Ferdinand-Jühlke-Straße 7
99095 Erfurt, Deutschland
produktsicherheit@kolibri360.de